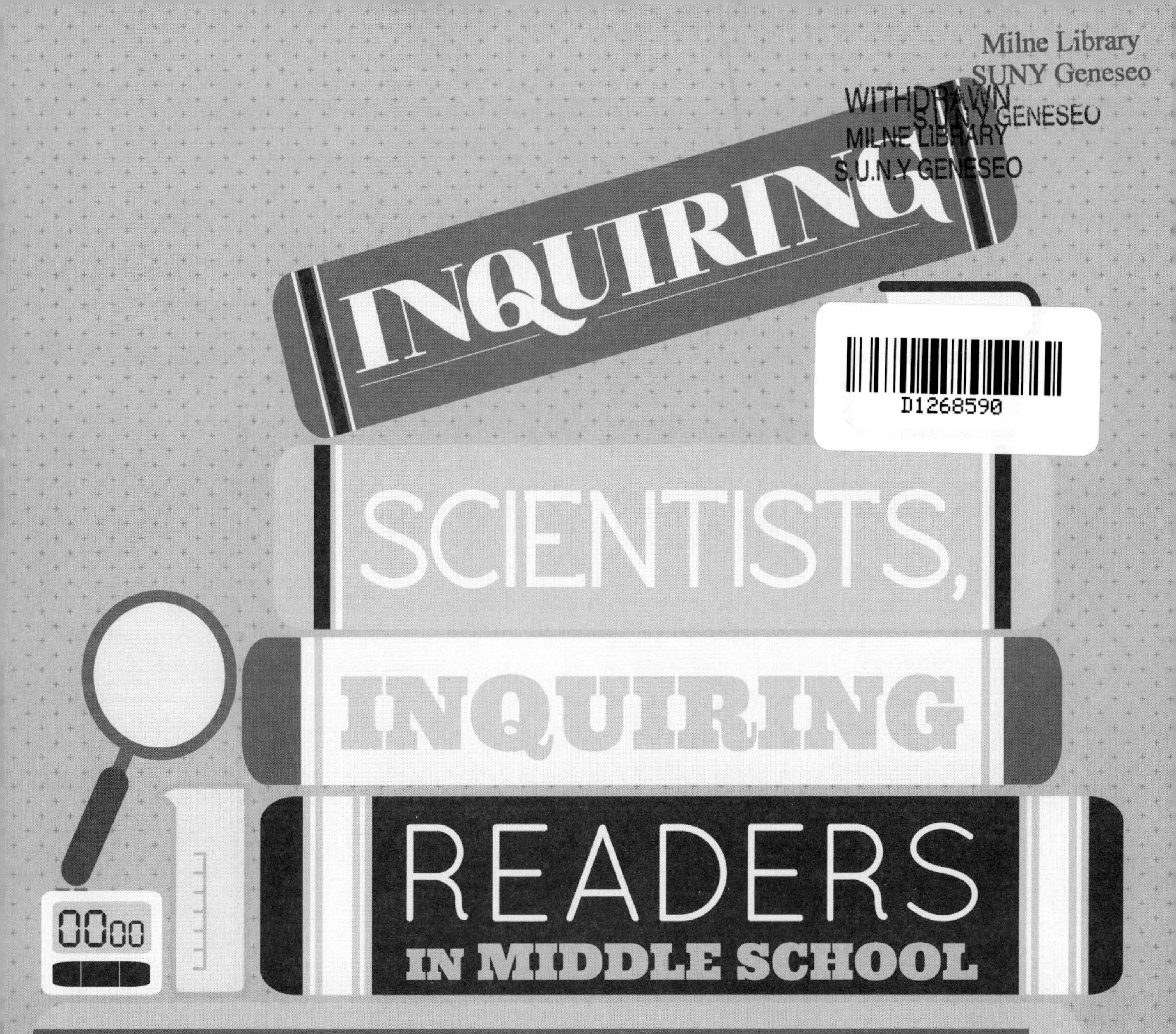


# Inquiring Scientists, Inquiring Readers in Middle School

## Using Nonfiction to Promote Science Literacy

GRADES 6–8

Shiverdecker &
Fries-Gaither

# Inquiring Scientists, Inquiring Readers in Middle School

## Using Nonfiction to Promote Science Literacy

### GRADES 6–8

Terry Shiverdecker
Jessica Fries-Gaither

Arlington, Virginia

Claire Reinburg, Director
Wendy Rubin, Managing Editor
Rachel Ledbetter, Associate Editor
Amanda Van Beuren, Associate Editor
Donna Yudkin, Book Acquisitions Coordinator

Art and Design
Will Thomas Jr., Director
Cover design by Rashad Muhammad and Joe Butera, Senior Graphic Designer
Interior design by Rashad Muhammad

Printing and Production
Catherine Lorrain, Director

National Science Teachers Association
David L. Evans, Executive Director
David Beacom, Publisher
1840 Wilson Blvd., Arlington, VA 22201
*www.nsta.org/store*
For customer service inquiries, please call 800-277-5300.


20 19 18 17 4 3 2 1



Cataloging-in-Publication Data for this book and the e-book are available from the Library of Congress.
ISBN: 978-1-68140-003-7
eISBN 978-1-68140-004-4

# CONTENTS

# CONTENTS

# ABOUT THE AUTHORS

Terry Shiverdecker is the K–12 STEM program director in the College of Education and Human Ecology at The Ohio State University in Columbus, Ohio. Terry works on developing and sustaining partnerships and pursuing state and national funding to support activities related to STEM education and teacher professional development. Previously, Terry was the director of STEM Initiatives at the Ohio Resource Center. She earned anf EdD from the University of Cincinnati, an MAT in Biology from Miami University, and a BS in Science Education from Wright State University. Terry has also taught high school science; taught as an adjunct instructor at the University of Cincinnati, the Lake Campus of Wright State University, and Ohio Wesleyan University; and served as the director of secondary curriculum and instruction at the Shelby County Educational Service Center.

Jessica Fries-Gaither is the lower school science specialist at the Columbus School for Girls in Columbus, Ohio, where she teaches science to grades 1–5. Prior to working at the Columbus School for Girls, she was an education resource specialist in the School of Teaching and Learning, College of Education and Human Ecology, at The Ohio State University. She served as the project director for two National Science Foundation–funded projects that supported elementary teachers in integrating science and literacy instruction. Jessica earned bachelor's degrees in the biological sciences and anthropology from the University of Notre Dame. She received her MEd from the University of Notre Dame's Alliance for Catholic Education program. She has taught middle school math and science in Memphis, Tennessee, and she has taught a variety of grade levels, ranging from eighth-grade science to fourth-grade self-contained, in Anchorage, Alaska.

# INTRODUCTION

Our journeys to understanding and, later, enthusiastically embracing and even advocating for the integration of literacy into inquiry-based science instruction took us down different paths. Jessica, a middle school science teacher with several years of elementary experience, first saw the importance of this integration while teaching fourth-grade students. She further developed this understanding while directing a National Science Foundation–funded grant to assist elementary teachers in using an integrated approach to teach science. For Terry, a former high school science teacher and science specialist, the journey didn't truly begin until she started viewing literacy through the lens of inquiry. Then, after reading "Science Text Sets: Using Various Genres to Promote Literacy and Inquiry" by Margaretha Ebbers (2002), she felt she had finally found a great example of how literacy can support science learning.

In her article, Ebbers describes how multiple genres of nonfiction (including biographies, field guides, reference books, and journals) present scientific information in different ways. Collections of these nonfiction works, known as text sets, can be used to support inquiry-based instruction. Finally, Terry thought, here was an example in which the literacy didn't overtake the need for rigorous, evidence-based inquiry science. As we planned a professional development workshop for elementary teachers that was based on this article, the idea for a book was born. Together, we published *Inquiring Scientists, Inquiring Readers: Using Nonfiction to Promote Science Literacy, Grades 3–5* (Fries-Gaither and Shiverdecker 2013).

Over the next few years, we presented our work at conferences and in other professional development settings. Conversations with teachers and curriculum specialists led us to realize that research-based, yet practical, information about integrating literacy was needed just as much at the middle school level. Thus, we began to write this book.

## ABOUT THIS BOOK

In this book, we share an approach to integrating all aspects of literacy (reading, writing, speaking, listening, and viewing) into inquiry-based science that is appropriate for middle school students. The units in Part II are inquiry-based and center on testable questions and data collection and analysis. These data are then used to make evidence-based conclusions. Each inquiry asks students to write. The writing ranges from scientific arguments to visual representations of a science concept. These practices are at the heart of the scientific enterprise and are highly engaging for students.

Each unit is constructed using the learning cycle framework. In this way, students are engaged in data collection and analysis before being expected to read to develop their understanding of the concept. This practice, often described by Terry as "experience before expertise," promotes critical thinking and enhances comprehension of the text. Reading may be used before inquiry to engage students or set the stage for the inquiry itself, but reading for content knowledge happens only after students have had a chance to develop their own understanding through the inquiry. Our approach is in direct contrast to the "read, then experiment to confirm" approach that many of us experienced in our own science education—an approach that is not nearly as engaging or effective in developing a rich understanding of science content.

The texts that students read, discuss, and write about are exclusively nonfiction and include children's literature, newspaper and magazine articles, articles published online, infographics, and even videos. Students engage in a wide variety of literacy activities before, during, and after reading to enhance their comprehension of the text and to link their inquiry experiences to what they have read. Only after these activities are completed are students expected to demonstrate mastery of concepts and vocabulary. Though this process takes time, the result is a deep and enduring understanding of science concepts.

We begin the book with a brief review of the research and best practices that serve as the foundation for our approach. First, we discuss the natural connections between science and literacy and review research that demonstrates that an integrated approach can increase student achievement in both areas. We then take a comprehensive look at the learning cycle, its history, and the myriad ways that inquiry and other science and engineering practices can be implemented into the classroom. Next, we turn our attention to literacy and how best practices can be applied to middle school science instruction. We also discuss the significant links between the *Next Generation Science Standards* (NGSS Lead States 2013) and the *Common Core State Standards, English Language Arts* (NGAC and CCSSO 2010) and the positive implications of those links for middle school teachers. Practical suggestions for getting started with the units conclude Part I of the book.

Part II consists of 10 units, and it is in those units that the research and best practices come to life. This is the heart of our work and where we know teachers will spend the most time. Each detailed unit includes scientific background information, common misconceptions that are associated with the content, an annotated list of the texts, safety considerations, strategies for differentiation, supporting documents (e.g., graphic organizers, writing prompts), and suggested assessments. The units encompass life, physical, and Earth and space science. All units have been tested in middle school science classrooms. An appendix provides information about grading and rubrics for evaluating student work products and science process skills.

## WHY THIS BOOK IS IMPORTANT

As science educators and professional development providers, we know that the current emphasis on literacy means that secondary science teachers need resources and support to meet current standards. We also know that today's students often arrive in middle school with specific needs in terms of reading and writing, which can frustrate teachers who have not received subject-specific guidance in meeting those needs. Again, support and resources are the keys. Finally, as lovers of science, we know that reading, writing, and communicating are part and parcel of the scientific enterprise. However, we strongly believe that any effort to integrate literacy into science instruction must do so while still preserving the quality of the science instruction itself. In our approach, literacy activities support the acquisition of science content through inquiry-based instruction. They do not replace active engagement with data with reading about science concepts.

Nonfiction text has unique demands that make it more difficult for students to read and comprehend. Without this understanding, teachers can mistakenly assume that students who do not complete assigned reading are simply failing to do homework. Many times, these students are experiencing difficulty with the text and give up. By incorporating reading strategies at appropriate points of a learning cycle unit, teachers can better support students and ensure that they are developing sufficient understanding through reading text.

Writing to communicate understanding is a key component of a scientist's work, yet students tend to be unfamiliar with the conventions of technical writing. Teaching students to write argument and other forms of scientific discourse better simulates the types of writing done in the scientific field. Of course, other forms of writing—quick writes to develop understanding and journaling about observations—have important places in the curriculum as well.

Scientific discourse is a crucial, yet regularly overlooked part of instruction. Too often, classroom conversation is best described as initiate-respond-evaluate, or IRE (Roller 1989). The teacher poses a question, the students answer, and the teacher informs the students if the response was correct or not. This pattern of discussion stems from a model of education in which a teacher's role is to transmit factual information rather than promote creativity and critical thinking (Beghetto 2013). Although this model has its place in the classroom at specific times, overreliance on it limits students' thinking and emphasizes correct answers over creativity and divergent thinking. Devoting time for students to talk to each other about what they are learning promotes deeper understanding and fulfills early adolescent students' developmental need for social engagement.

## HOW TO USE THIS BOOK

Although we have positioned the research background as the first section in our book, we know that many classroom teachers will gravitate toward the units. In fact, we admit that we would be tempted to do the same if handed a copy of this book! Our research into reading nonfiction tells us that having an authentic context is vital for comprehension. Beginning with the units does, in fact, provide that context. So if you're inclined to begin with Part II, feel free to start there. But please do go back and spend time with the research review. Education is a research-based profession. Research findings inform our practice and help us better meet the needs of our students. You will significantly enhance your understanding of the units if you take the time to read these first few chapters.

Embedding nonfiction text and literacy activities into inquiry-based science honors the best practices of both disciplines. We began this introduction by sharing our journeys to not only understanding but also advocating for an integrated approach. We hope that our work helps you along your own path.

## REFERENCES

Beghetto, R. A. 2013. *Killing ideas softly? The promise and perils of creativity in the classroom.* Charlotte, NC: Information Age Publishing.

Ebbers, M. 2002. Science text sets: Using various genres to promote literacy and inquiry. *Language Arts* 80 (1): 40–50.

Fries-Gaither, J., and T. Shiverdecker. 2013. *Inquiring scientists, inquiring readers: Using nonfiction to promote science literacy, grades 3–5.* Arlington, VA: NSTA Press.

National Governors Association Center for Best Practices and Council of Chief State School Officers (NGAC and CCSSO). 2010. *Common core state standards.* Washington, DC: NGAC and CCSSO.

NGSS Lead States. 2013. *Next Generation Science Standards: For states, by states.* Washington, DC: National Academies Press. *www.nextgenscience.org/next-generation-science-standards.*

Roller, C. 1989. Classroom interaction patterns: Reflections of a stratified society. *Language Arts* 66 (5): 492–500.

Part I
INTEGRATING SCIENCE
AND
LITERACY INSTRUCTION

# Chapter 1
# Literacy in Support of Science Learning

Science educators may be very tempted to relinquish responsibility for literacy instruction to someone else—the English language arts teacher, for instance. After all, we have *a lot* of science to teach. Nestled within all of the science we have to teach are things such as diagrams, charts, graphs, formulas, lab reports, maps, micrographs … the list goes on and on. Deciphering all of these things is *reading;* constructing them is *writing.* Literacy is as fundamental to science as a microscope, test tube, graduated cylinder, or balance. We cannot *do* science without it. We must use literacy skills to engage in the practice of science. Why would we relinquish responsibility for literacy instruction when it is so intimately intertwined with inquiry-based science instruction?

Literacy can be used to support science learning in a multitude of ways: linking what is revealed through firsthand investigations to generally accepted scientific knowledge (Fang, Lamme, and Pringle 2010), gathering information through secondhand investigations (Pearson, Moje, and Greenleaf 2010), developing vocabulary in a meaningful way (Fang et al. 2008), illustrating science-specific ways of talking and writing (Pearson, Moje, and Greenleaf 2010), and organizing thinking prior to writing. Just as literacy can be used to build science knowledge and skills, we can use inquiry to support the development of literacy skills. Inquiry-based science can also be used to support literacy instruction. Krajcik and Sutherland (2010) have identified five aspects of literacy that can be developed through inquiry-based science, as follows:

1. Linking new ideas to prior knowledge and experiences
2. Anchoring learning in questions that are important in the lives of students
3. Connecting multiple representations
4. Providing opportunities for students to use science ideas
5. Supporting students' engagement with the discourses of science

The overall benefit of integration is increased student achievement in both science and literacy (Fang et al 2008; Romance and Vitale 1992).

The benefits of integrating literacy into science instruction are clear, but there are some challenges. One of the challenges associated with integration is the fear that literacy will overshadow inquiry—instead of doing science, students will be reading about it. Other challenges include the poor quality of available texts, students' struggles with expository text, and secondary teachers who lack preparation in literacy instruction (Pearson, Moje, and Greenleaf 2010). It is also likely that secondary teachers do not feel they have time to integrate literacy instruction. Incorporating literacy instruction doesn't mean secondary teachers will be helping students learn to read at the most basic level (e.g., decoding text, developing phonemic awareness), nor does it mean they will be teaching students writing skills such as basic sentence structure. It does mean that teachers will arm students with strategies for interpreting, communicating, and making meaning of the many ways we represent science.

In short, they will be arming students with strategies for reading and writing to learn about science. This includes explicit instruction on literacy strategies. Strategy instruction leads to gains in science knowledge and literacy skills (Radcliffe et al. 2008). Literacy strategies help students make meaning of the many ways science is communicated.

## INTEGRATION EXAMPLES

Think about each of the following examples in the context of an inquiry investigation in which students discover that plants gain mass from atmospheric carbon, not from soil or water as they may have predicted. They have not yet discovered the process of photosynthesis and are trying to make sense of how plants can take carbon from the atmosphere.

### Linking Inquiry-Based Knowledge to New Learning Through Nonfiction Text

The process of photosynthesis is difficult to uncover through firsthand inquiry investigations. In this case, nonfiction text can be used to build on what students learned through their earlier investigations. It is likely that the reading will be challenging and will include scientific terminology, diagrams of cellular structures, and a chemical equation. The objective is for the students to connect this new information to what they learned through their firsthand investigation. Ultimately, they will construct a cohesive understanding of photosynthesis from the macroscopic to microscopic level. Making meaning from the complex text and connecting it to previous knowledge is a high-level cognitive task. For students to meet the objective, they need as much support for the literacy demands as they do for the content demands. Support can be provided through literacy strategy instruction, followed by teacher modeling, group practice, and then individual practice (Duke and Pearson 2002; Pearson and Gallagher 1983).

Following an investigation with purposeful reading from nonfiction texts can help students connect what they learned through the firsthand investigation with current scientific knowledge. The firsthand investigation gives them an authentic experience that they can link with the established content knowledge. Reading to develop content knowledge before engaging in an authentic experience often results in the "I read it but I didn't understand it" phenomenon. Students need experiences that they can build on as they read. Combining the inquiry investigation and the nonfiction reading helps students construct the desired science knowledge. Students gain experience and then build expertise.

## Using Secondhand Investigations

Beyond the nonfiction texts, analogies, simulations, videos, or other resources could be used in the secondhand investigation of photosynthesis. All of these nonfiction resources can play a critical role in helping students construct meaning for concepts they cannot explore through firsthand investigations. At first blush, this may seem like we have abandoned *doing* science in favor of *reading about* science—one of secondary teachers' fears about integrating literacy. Nothing could be further from the truth. In secondhand investigations, students use nonfiction resources as sources of data as they gather information in response to carefully crafted questions (Pearson, Moje, and Greenleaf 2010). Students extract information from the diagrams, data sets, charts, graphs, maps, equations, formulas, micrographs, and so on that accompany the text, as well as from the text itself. This information serves as the evidence that will be used to make a claim and construct a logical argument. Using text and other resources as the foundation for inquiry is vastly different from using text to look up an answer.

## Engaging in the Discourse of Science

Discourse around photosynthesis, even at the middle school level, is likely to include scientific terminology (e.g., chloroplasts, chlorophyll, stroma, carbohydrate), a chemical equation ($6 H_2O + 6 CO_2 \rightarrow C_6H_{12}O_6 + 6 O_2$), and at least one diagram (Figure 1.1 shows several simple photosynthesis diagrams). Each of these components by itself may provide sufficient information about photosynthesis. But when they are combined, a more complete picture of a complex topic emerges. Combined, they are a fair representation of how we communicate in science. The literacy demands associated with synthesizing all of these representations into a cohesive understanding of photosynthesis is very high. As stated earlier, students need support in literacy as well as content to make meaning of these representations and connect them to prior investigations.

Science terminology is an integral part of learning and communicating about science. Dealing with the terminology may be one of the most frustrating aspects of teaching science. Studies

**FIGURE 1.1. SAMPLE PHOTOSYNTHESIS DIAGRAMS**

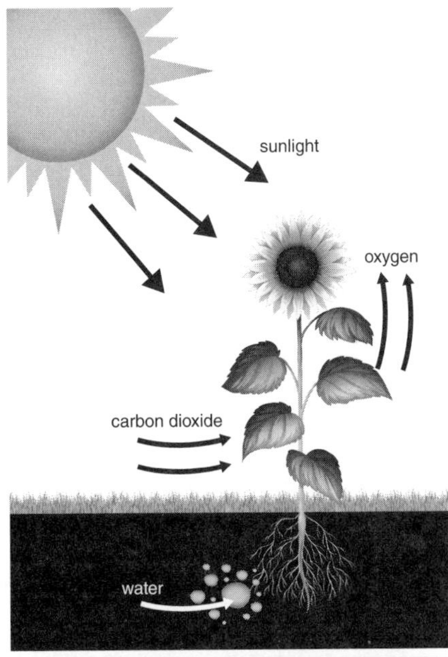


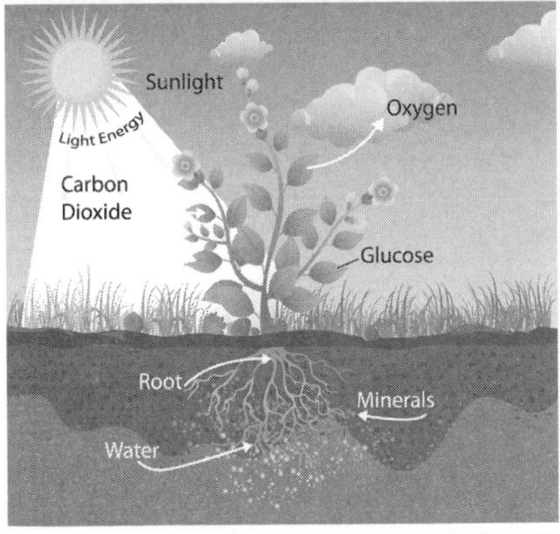


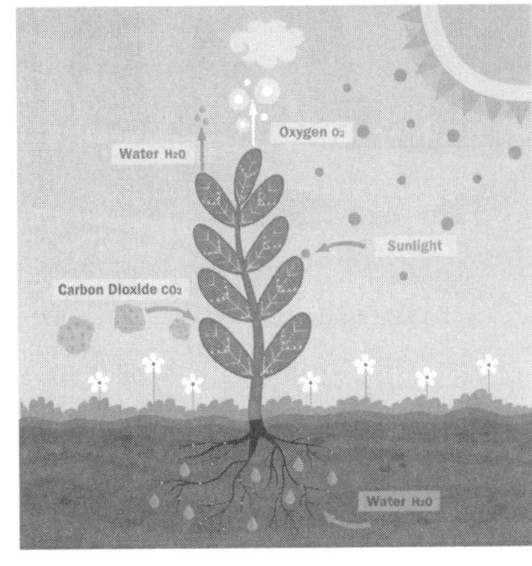

have shown (Groves 1995; Yager 1983) that a typical high school science course introduces more new words than a typical high school foreign language course. The overemphasis on vocabulary supports the misconception that science is a set of unrelated facts. There are several approaches to using literacy in support of science learning that can alleviate the problem of too much vocabulary: Purposefully select nonfiction texts that limit science terminology to the terms necessary to understand the concept, use a just-in-time approach to introduce new vocabulary, combine evidence from investigations with evidence from texts to create student-constructed definitions, and use graphic organizers to help students construct a definition and situate the term in the context of the concept.

## CONCLUSION

Literacy is central to our ability to communicate what we know about science. We talk and write about science differently than we talk and write about other subjects (Gee 2004; Varelas et al. 2008). In fact, the way we use language in science can be a stumbling block for students. Along with all of the ways to represent science knowledge mentioned earlier, the way we use language in science is very precise. In common language, we might describe a dandelion as a bright yellow flower commonly found in lawns and typically thought of as a weed. Kansas State University's Department of Horticulture, Forestry, and Recreation Resources (2014) describes the dandelion as "a *perennial* herb that forms a *rosette* in lawns and gardens. *Inflorescence* composed of yellow ray *florets* … give rise to a 'puffball' head. New plants germinate primarily in the fall (late September)." From the scientist's perspective, this is a fairly short description of the dandelion. For most students, this precise use of language, along with the multitude of nonlinguistic ways we represent science, is a very foreign way to communicate. The only way students can learn the language of science is to be immersed in it and to read, write, speak, and listen to it daily. In other words, language skills are built by paying attention to the literacy demands that come with learning science.

## REFERENCES

Duke, N. K., and P. D. Pearson. 2002. Effective practices for developing reading comprehension. In *What research has to say about reading instruction, 3rd ed.*, ed. A. E. Farstrup and S. J. Samuels, 205–242. Newark, DE: International Reading Association.

Fang, Z., L. Lamme, and R. Pringle. 2010. *Language and literacy in inquiry-based science classrooms, grades 3–8.* Thousand Oaks, CA: Corwin Press.

Fang, Z., L. Lamme, R. Pringle, J. Patrick, J. Sanders, C. Zmach, S. Charbonnet, and M. Henkel. 2008. Integrating reading into middle school science: What we did, found and learned. *International Journal of Science Education* 30 (15): 2067–2089.

Gee, J. P. 2004. Language in the science classroom: Academic social languages as the heart of school-based literacy. In *Crossing borders in literacy and science instruction,* ed. E. W. Saul, 13–32. Newark, DE: International Reading Association.

Groves, F. H. 1995. Science vocabulary load of selected secondary science textbooks. *School Science and Mathematics* 95 (5): 231–235.

Kansas State University, Department of Horticulture, Forestry, and Recreation Resources. 2014. Problem: Dandelion *(Taraxacum officinale). www.hfrr.ksu.edu/doc2032.ashx.*

Krajcik, J. S., and L. M. Sutherland. 2010. Supporting students in developing literacy in science. *Science* 328 (456): 456–459.

Pearson, P. D., and M. C. Gallagher. 1983. The instruction of reading comprehension. *Contemporary Educational Psychology* 8 (3): 317–344.

Pearson, P. D., E. Moje, and C. Greenleaf. 2010. Literacy and science: Each in the service of the other. *Science* 328 (456): 459–463.

Radcliffe, R., D. Caverly, J. Hand, and D. Franke. 2008. Improving reading in a middle school science classroom. *Journal of Adolescent and Adult Literacy* 51 (5): 398–408.

Romance, N. R., and M. R. Vitale. 1992. A curriculum strategy that expands time for in-depth elementary science instruction by using science-based reading strategies: Effects of a year-long study in grade four. *Journal of Research in Science Teaching* 29 (6): 545–554.

Varelas, M., C. C. Pappas, J. M. Kane, A. Arsenault, J. Hankes, and B. M. Cowan. 2008. Urban primary-grade children think and talk science: Curricular and instructional practices that nurture participation and argumentation. *Science Education* 92 (1): 65–95.

Yager, R. E. 1983. The importance of terminology in teaching K–12 science. *Journal of Research in Science Teaching* 20 (6): 577–588.

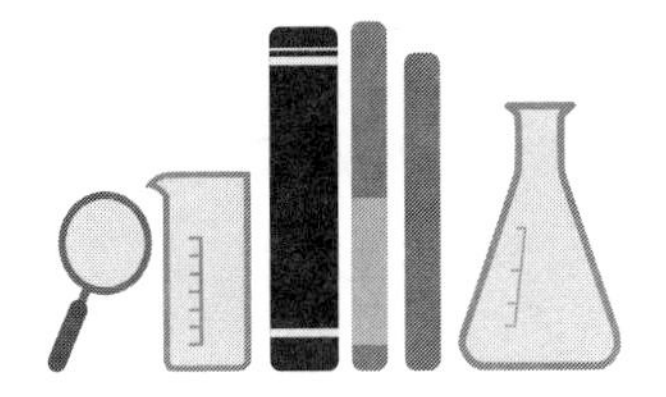

**Chapter 2**

# Science and Engineering Practices and the Learning Cycle

Twelve-year-old boys have a special knack for bringing fear to the hearts of their mothers when they ask, "Mom, where's the duct tape?" Such is the case with my grandson. My daughter has learned not to ask Ethan what he wants the duct tape (or glue or stapler or whatever else) for when he asks. She just shakes her head and hopes for the best. On a particular evening in March, the best is what she got.

Ethan had decided to make a bow and arrow out of bamboo skewers, rubber bands, and (of course) duct tape. He had no particular need for a bamboo skewer bow and arrow, he just wanted to know if it was possible to make one. He surveyed the materials available to him, tested their properties, and proceeded with his plan. After a few false starts and quick adjustments, he had made a bamboo skewer bow and arrow that was surprisingly accurate. (He had also made a target out of a shoe box.) By the end of the night, he was able to drop and roll past the target and still hit the bull's-eye!

What Ethan (and my daughter) didn't know was that he was facing a design challenge. Like any good engineer, he went to work solving the problem with the limitations he faced. His data were anecdotal, but they were sufficient for him to know how to redesign his bow. That bamboo-skewer bow and arrow is long gone, but the ability to ask questions or identify problems and go about finding solutions remains. It will undoubtedly serve Ethan well for years to come.

—Terry

## THE LEARNING CYCLE

The learning-cycle framework has been an important tool in planning science instruction for more than 50 years. Robert Karplus, a physicist, developed the framework after visiting his daughter's second-grade class to discuss electricity and thinking about the ways in which children learn science. The original learning-cycle framework consisted of three phases: Exploration, Invention, and Discovery (Karplus and Fuller 2002). As research in the field of education contributed to a richer understanding of learning, the cycle was revised. One major and well-known revision is the Biological Sciences Curriculum Study (BSCS) 5E approach: Engagement, Exploration, Explanation, Elaboration, and Evaluation (Bybee et al. 2006). This version built on Karplus's original model, adding the Engagement phase in which the teacher guides students in accessing prior knowledge and adding a phase to evaluate student learning. A similar 4E approach consists of Exploration, Explanation, Expansion, and Evaluation phases (Martin et al. 1997). One major difference between the BSCS 5E approach and Martin's 4E model is that in Martin's model, the Evaluation phase is ongoing throughout the framework to assess skills and content knowledge. Other versions include a 7E approach that adds Elicit and Extend phases (Eisenkraft 2003) and a 6E approach that integrates technology in an ongoing E-Search phase through the entire framework (Chessin and Moore 2004).

## OUR APPROACH: ENGAGE, EXPLORE, EXPLAIN, EXPAND, ASSESS

Our learning-cycle framework is like the BSCS 5E model, but it includes Martin's concept of ongoing assessment. In fact, we prefer to use the term *assess* rather than *evaluate* to better reflect the importance of ongoing, varied, and embedded assessment. Figure 2.1 illustrates our learning-cycle framework. An overview of each phase follows.

### FIGURE 2.1. OUR APPROACH TO THE LEARNING CYCLE

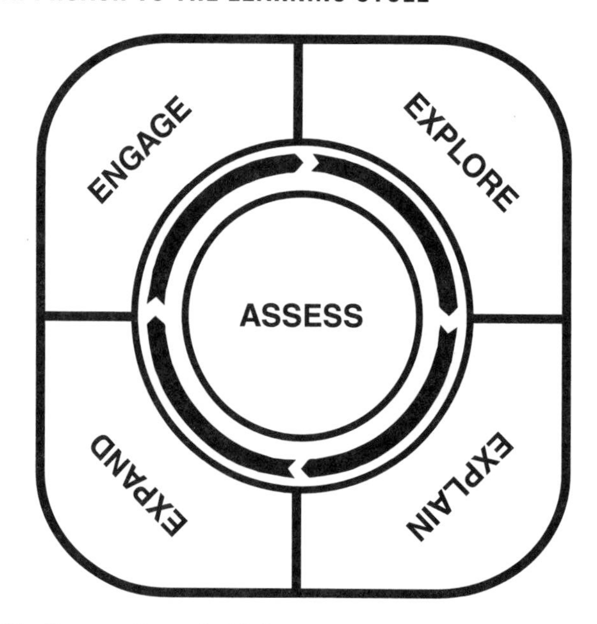


*Source:* Adapted from Ohio Resource Center for Mathematics, Science, and Reading 2012.

***Engage.*** The purpose of the Engage phase is to access prior knowledge, expose misconceptions, build interest and enthusiasm for the topic, and instill in students a need to know more about the topic. A wide variety of activities can serve this purpose, including discrepant events, demonstrations, read-alouds, or independent reading of captivating text. It is important to note that although text or videos may be used in the Engage phase, the purpose is to interest students, not to provide instruction in the science content to be developed later. In our units, Engage phase activities range from watching video clips of the aurora to comparing the mass of a seed to the mass of a tree. The Engage phase also introduces the testable question that is investigated over the course of the unit.

***Explore.*** The Explore phase is dedicated to active investigation. Students are working to answer the testable question that was introduced in the Engage phase. Science practices are nurtured as students make observations, collect data, develop graphs, make predictions, and draw conclusions through first- or secondhand investigations. Vocabulary is introduced in a just-in-time fashion so that terms are provided only when students are ready for them. Students now have a meaningful context in which to place the vocabulary. It is important to remember that students will not have mastered vocabulary or concepts at this time. In our inquiry units, students engage in a wide range of experiences that build science practices while revealing the concept being studied.

***Explain.*** In this phase, students refer to evidence from their investigations and use scientifically accurate terminology as they share what they have learned. A common misconception about the Explain phase is that it is the teacher's responsibility to explain the correct concepts; rather, the responsibility to explain lies with the students. The Explain phase often involves the creation of a product that showcases student learning. Some of the products students develop in the Explain phase of our inquiry units include scientific arguments, infographics, explanatory reports, blog posts, and other forms of written and visual work.

***Expand.*** The Expand phase provides opportunities to situate new learning in real-world contexts, apply knowledge in a design challenge, begin investigating a related concept, or delve deeper into the current concept. The goal of the Expand phase is to deepen knowledge. In our inquiry units, students deepen and extend their knowledge in various ways, such as designing a reusable container to keep hot meals hot and cold meals cold, using their understanding of cell organelles to determine if *Euglena* are plants or animals, researching inherited diseases and disorders and presenting their findings at a mock medical conference, and investigating the effect of various wavelengths of light on photosynthesis.

***Assess.*** The Assess phase is embedded throughout the learning cycle. This allows for both formative and summative assessment. Formative assessments are typical in the Engage and Explore phases. The Explain phase always has a summative assessment but can also include a formative assessment. The expand phase may include either formative or summative assessments, depending on the nature of the Expand activities. Some of the formative assessments used in our inquiry units are guiding questions, observations of students at work, graphic organizers, and quick writes.

It is important to remember that the learning cycle is a *framework,* not a rigid mold into which a lesson or unit must fit. The framework can be used flexibly to fit the needs of teacher, students, and content at hand. For example, a unit that covers a large amount of content might have multiple Explore and Explain phases so that students can master each concept before moving on to the next. Multiple learning cycles can also be linked together into a coherent sequence of instruction. This might be done by closely connecting the Expand phase of one unit with the Engage phase of the next. In other cases, the Expand phase itself might serve as the Engage phase for the subsequent unit. The units presented in Chapters 14 and 15 are linked in this way. When concepts are connected to each other through a repeated learning-cycle framework, students are able to place concepts in a meaningful context, rather than learn facts in isolation.

Strengths and Limitations of the Learning Cycle The learning cycle is recognized in science education as an outstanding model for planning inquiry instruction. A robust body of research indicates that science instruction using the Science Curriculum Improvement Study and 5E learning cycles result in the development of more sophisticated scientific reasoning and an increased interest in science content. Moreover, the 5E approach leads to increased mastery of the content (Bybee et al. 2006).

Brown and Abell (2007) found that hands-on activities are a necessary but insufficient means of teaching science. If teachers do not work through an entire learning cycle with their students, student learning suffers. Roth (as cited in Brown and Abell 2007) found that fifth-grade students understood scientific concepts better when the Exploration phase was followed with discussion and writing. The discussion and writing occur in the Explain phase, after the hands-on experiences common to the Explore phase. There is sometimes a tendency to rush through learning-cycle lessons or units. Skipping phases or giving in to the temptation to explain the concepts to students during the Explore phase results in less satisfactory results. Good science instruction, particularly when it includes writing and the development of products, takes time.

Another limitation arises when instructional planning focuses on identifying *E* activities without careful consideration of the learning that should be taking place in each phase. The Es in the 5E learning cycle are easy to remember. But we've seen too many "learning cycle" lessons that were nothing more than a series of hands-on activities. For the learning cycle to be effective, the activities must be a coherent, carefully structured set of experiences that allows students to construct meaning by linking evidence and explanations.

## SCIENCE AND ENGINEERING PRACTICES AND THE LEARNING CYCLE

The learning cycle is an excellent framework for the development of the science and engineering practices set forth in *A Framework for K–12 Science Education: Practices, Crosscutting Concepts, and Core Ideas* (NRC 2012):

1. Asking questions (for science) and defining problems (for engineering)
2. Developing and using models
3. Planning and carrying out investigations
4. Analyzing and interpreting data
5. Using mathematics and computational thinking
6. Constructing explanations (for science) and designing solutions (for engineering)

7. Engaging in argument from evidence
8. Obtaining, evaluating, and communicating information

The focus on a guided investigation and the answering of a testable question throughout the learning cycle naturally leads to these types of activities. But what exactly is meant by each of these practices, and where do they fit within the context of the learning cycle (see Tables 2.1–2.8)?

## Asking Questions and Defining Problems

### TABLE 2.1. ASKING QUESTIONS AND DEFINING PROBLEMS

| Science | Engineering |
| --- | --- |
| Science begins with a question about a phenomenon, such as "Why is the sky blue?" or "What causes cancer?" A basic practice of the scientist is the ability to formulate empirically answerable questions about phenomena to establish what is already known, and to determine what questions have yet to be satisfactorily answered. | Engineering begins with a problem that needs to be solved, such as "How can we reduce the nation's dependence on fossil fuels?" or "What can be done to reduce a particular disease?" or "How can we improve the fuel efficiency of automobiles?" A basic practice of engineers is to ask questions to clarify the problem, determine criteria for a successful solution, and identify constraints. |

*Source:* Bybee 2011.

The learning cycle provides multiple opportunities to ask questions. Broad and overarching questions, or essential questions, can be posed during the Engage phase and answered through repeating Explore/Explain cycles. In this scenario, each Explore/Explain cycle addresses a subset of the knowledge needed to fully answer the question. Additionally, more focused questions may arise in the Explore phase while an investigation is already underway. These questions may help students clarify their thinking or make connections between their investigation and observations they have made beyond the classroom. Finally, novel questions that build on the knowledge gained through investigation can be posed in the Expand phase.

During a learning cycle, both the teacher and the students can generate questions. Teacher-generated questions can direct the unit, assess student understanding, and challenge students to apply their knowledge to novel situations. For students to truly engage in this practice, they need to have time to do so. The frequent opportunities for dialogue in our inquiry units provide chances for reflection and posing questions.

## Developing and Using Models

### TABLE 2.2. DEVELOPING AND USING MODELS

| Science | Engineering |
|---|---|
| Science often involves the construction and use of models and simulations to help develop explanations about natural phenomena. Models make it possible to go beyond observables and simulate a world not yet seen. Models enable predictions of the form "if ... then ... therefore" to be made in order to test hypothetical explanations. | Engineering makes use of models and simulations to analyze extant systems to identify flaws that might occur, or to test possible solutions to a new problem. Engineers design and use models of various sorts to test proposed systems and to recognize the strengths and limitations of their designs. |

*Source:* Bybee 2011.

This practice refers to more than the construction of physical models. Abstract models, or testable ideas, are widely used as students analyze data and make sense of phenomena. Developing and using models occurs across the learning cycle during the Explore, Explain, and Expand phases. Data collected in the Explore phase can be used to construct and test the model. In the Explain phase, students use the model to predict what will happen under a given set of conditions. In the Expand phase, students apply the model to a new and unique situation. Several of the inquiry units found in Part II involve the creation of models. In Chapter 6, "Modeling Cells," students not only create physical models of plant and animal cells, but also use a mental model to predict whether *Euglena* are plants or animals. In Chapter 10, "Nature's Light Show: It's Magnetic!," students use a mental model to predict whether planets would experience an aurora.

## Planning and Carrying Out Investigations

### TABLE 2.3. PLANNING AND CARRYING OUT INVESTIGATIONS

| Science | Engineering |
|---|---|
| Scientific investigations may be conducted in the field or in the laboratory. A major practice of scientists is planning and carrying out systematic investigations that require clarifying what counts as data and identifying variables in experiments. | Engineering investigations are conducted to gain data essential for specifying criteria or parameters and to test proposed designs. Like scientists, engineers must identify relevant variables, decide how they will be measured, and collect data for analysis. Their investigations help them to identify the effectiveness, efficiency, and durability of designs under different conditions. |

*Source:* Bybee 2011.

Students most often plan and carry out investigations in the Explore phase of the learning cycle. It is not unusual and is, in fact, acceptable if these initial investigations are teacher guided. Students can then plan and carry out a student-led investigation in the Expand phase. In Chapter 8, "Seriously … That's Where the Mass of a Tree Comes From?" students design and conduct investigations about the effects of various wavelengths of light on photosynthesis.

## Analyzing and Interpreting Data

### TABLE 2.4. ANALYZING AND INTERPRETING DATA

| Science | Engineering |
|---|---|
| Scientific investigations produce data that must be analyzed in order to derive meaning. Because data usually do not speak for themselves, scientists use a range of tools—including tabulation, graphical interpretation, visualization, and statistical analysis—to identify the significant features and patterns in the data. Sources of error are identified and the degree of certainty calculated. Modern technology makes the collection of large data sets much easier, providing secondary sources for analysis. | Engineering investigations include analysis of data collected in the tests of designs. This allows comparison of different solutions and determines how well each meets specific design criteria—that is, which design best solves the problem within given constraints. Like scientists, engineers require a range of tools to identify the major patterns and interpret the results. Advances in science make analysis of proposed solutions more efficient and effective. |

*Source:* Bybee 2011.

Data analysis can occur at any point in the learning cycle. In the Engage phase, an analysis of a small set of data can pique students' interest and instill a need to know more. In the Explore phase, data analysis follows the collection of data that typically occurs. In a well-designed investigation, students will collect data from multiple trials, which allows trends to emerge. In the Explain phase, students may analyze and interpret data to demonstrate their knowledge of the phenomenon being studied. Data analysis and interpretation in the Expand phase should build on previous work and be more sophisticated and complex. Firsthand data can be collected with traditional tools or with electronic probes if they are available. Secondhand data can be collected through simulations and online data sets. All our inquiry units provide opportunities for data collection. Chapter 11, "Thermal Energy: An Ice Cube's Kryptonite!" is one example of firsthand data collection, whereas Chapter 15, "The Toes and Teeth of Horses," is an example of inquiry using secondhand data.

## Using Mathematics and Computational Thinking

### TABLE 2.5. USING MATHEMATICS AND COMPUTATIONAL THINKING

| Science | Engineering |
|---|---|
| In science, mathematics and computation are fundamental tools for representing physical variables and their relationships. They are used for a range of tasks such as constructing simulations; statistically analyzing data; and recognizing, expressing, and applying quantitative relationships. Mathematical and computational approaches enable prediction of the behavior of physical systems along with the testing of such predictions. Moreover, statistical techniques are also invaluable for identifying significant patterns and establishing correlational relationships. | In engineering, mathematical and computational representations of established relationships and principles are an integral part of the design process. For example, structural engineers create mathematical-based analysis of designs to calculate whether they can stand up to expected stresses of use and if they can be completed within acceptable budgets. Moreover, simulations provide an effective test bed for the development of designs as proposed solutions to problems and their improvement, if required. |

*Source:* Bybee 2011.

In a learning-cycle approach, mathematics and computational thinking are an integral part of modeling (when appropriate for the content) and data interpretation and analysis. Consequently, students are most likely to engage in mathematics and computational thinking when they are engaged in the other practices. For example, students calculate the predicted frequency of different traits in Chapter 7, "The Genetic Game of Life," and calculate the density of various substances in Chapter 9, "Chemistry, Toys, and Accidental Inventions."

## Constructing Explanations and Designing Solutions

### TABLE 2.6. CONSTRUCTING EXPLANATIONS AND DESIGNING SOLUTIONS

| Science | Engineering |
|---|---|
| The goal of science is the construction of theories that provide explanatory accounts of the material world. A theory becomes accepted when it has multiple independent lines of empirical evidence, greater explanatory power, a breadth of phenomena it accounts for, and has explanatory coherence and parsimony. | The goal of engineering design is a systematic solution to problems that is based on scientific knowledge and models of the material world. Each proposed solution results from a process of balancing competing criteria of desired functions, technical feasibility, cost, safety, aesthetics, and compliance with legal requirements. Usually there is no one best solution, but rather a range of solutions. The optimal choice depends on how well the proposed solution meets criteria and constraints. |

*Source:* Bybee 2011.

Constructing explanations is an ongoing cognitive task that students begin in the Engage phase. Developing a working knowledge of scientific phenomena takes time and is fostered by rich and repeated experiences. The experiences occur throughout the learning cycle as students delve deeper into the content. Each phase of the learning cycle offers opportunities for students to construct, tear down, and rebuild their mental representations of what they are investigating. This practice takes center stage in all of our inquiry units, although the context and format in which students construct these explanations varies depending on the topic at hand.

## Engaging in Argument From Evidence

### TABLE 2.7. ENGAGING IN ARGUMENT FROM EVIDENCE

| Science | Engineering |
|---|---|
| In science, reasoning and argument are essential for clarifying strengths and weaknesses of a line of evidence and for identifying the best explanation for a natural phenomenon. Scientists must defend their explanations, formulate evidence based on a solid foundation of data, examine their understanding in light of the evidence and comments by others, and collaborate with peers in searching for the best explanation for the phenomena being investigated. | In engineering, reasoning and argument are essential for finding the best solution to a problem. Engineers collaborate with their peers throughout the design process, with a critical stage being the selection of the most promising solution among a field of competing ideas. Engineers use systematic methods to compare alternatives, formulate evidence based on test data, make arguments to defend their conclusions, critically evaluate the ideas of others, and revise their designs in order to identify the best solution. |

*Source:* Bybee 2011.

Argumentation most often occurs in the Explain and Expand phases of the learning cycle. Students gather data in the Explore phase, then use this data to construct an argument in the Explain phase. Follow-up investigations in the Expand phase produce additional data that may strengthen students' arguments or may cause them to revise their thinking. Students engage in argumentation in Chapter 12, "Landfill Recovery," and in Chapter 13, "Sunlight and the Seasons."

## Obtaining, Evaluating, and Communicating Information

### TABLE 2.8. OBTAINING, EVALUATING, AND COMMUNICATING INFORMATION

| Science | Engineering |
|---|---|
| Science cannot advance if scientists are unable to communicate their findings clearly and persuasively or learn about the findings of others. A major practice of science is thus to communicate ideas and the results of inquiry—orally; in writing; with the use of tables, diagrams, graphs and equations; and by engaging in extended discussions with peers. Science requires the ability to derive meaning from scientific texts such as papers, the internet, symposia, or lectures to evaluate the scientific validity of the information thus acquired and to integrate that information into proposed explanations. | Engineering cannot produce new or improved technologies if the advantages of their designs are not communicated clearly and persuasively. Engineers need to be able to express their ideas orally and in writing; with the use of tables, graphs, drawings or models; and by engaging in extended discussions with peers. Moreover, as with scientists, they need to be able to derive meaning from colleagues' texts, evaluate information, and apply that information usefully. |

*Source:* Bybee 2011.

Students have numerous opportunities throughout the learning cycle to obtain, evaluate, and communicate information. This is especially true when nonfiction text is seamlessly integrated into the inquiry process. In the Engage phase, student communications most often consist of discussions in response to an intriguing stimulus. They are likely to obtain information from a variety of media, such as videos, nonfiction text, infographics, and so on. In the Explore phase, information is often obtained through firsthand investigations and supplemented with nonfiction texts. Students communicate in a variety of ways in the Explain phase. Communication can include reports, diagrams and drawings, graphical representations of data, stories, and writing prompts. Finally, in the Expand phase, students gather additional information as they did in the Explore phase, but now the information and communication are more sophisticated and complex. This is another practice woven throughout all of our inquiry units in various ways.

## BUT WHAT ABOUT PROCESS SKILLS?

Teachers may wonder why the recent literature has shifted from focusing on scientific process skills (Padilla 1990) to science and engineering practices. The authors of *A Framework for K–12 Science Education* explain that they "use the term 'practices' instead of a term such as 'skills' to emphasize that engaging in scientific investigation requires not only skill but also knowledge that is specific to each practice" (NRC 2012, p. 30). Process skills, such as observing, inferring, and defining operationally, still have important roles to play in science classrooms.

Table 2.9 (p. 24) illustrates the relationship between the science and engineering practices and the process skills. In fact, you can't engage in the science and engineering practices without the process skills. Instead, the new language reflects a belief that many science educators have held all along: that these behaviors and habits of mind are best taught within the context of science content. While the science and engineering practices are woven throughout our inquiry units, we have also provided rubrics for assessing science process skills in Appendix 1. We believe that teachers need, at different times, to move flexibly between practices and process skills as they plan, implement, and assess the effectiveness of inquiry-based instruction.

### TABLE 2.9. CORRELATION BETWEEN SCIENCE AND ENGINEERING PRACTICES AND THE PROCESS SKILLS

| Science and Engineering Practices | Supporting Process Skills |
|---|---|
| Asking questions (for science) and defining problems (for engineering) | • Communicating<br>• Observing |
| Developing and using models | • Communicating<br>• Formulating models<br>• Observing<br>• Predicting |
| Planning and carrying out investigations | • Controlling variables<br>• Defining operationally<br>• Experimenting<br>• Formulating hypotheses<br>• Inferring<br>• Measuring<br>• Observing<br>• Predicting |
| Analyzing and interpreting data | • Controlling variables<br>• Defining operationally<br>• Experimenting<br>• Formulating hypotheses<br>• Inferring<br>• Measuring<br>• Observing<br>• Predicting |
| Using mathematics and computational thinking | • Predicting<br>• Interpreting data<br>• Formulating hypotheses<br>• Formulating models |
| Constructing explanations (for science) and designing solutions (for engineering) | • Classifying<br>• Communicating<br>• Formulating models<br>• Inferring<br>• Interpreting data<br>• Observing<br>• Predicting |
| Engaging in argument from evidence | • Communicating<br>• Interpreting data<br>• Inferring<br>• Predicting |
| Obtaining, evaluating, and communicating information | • Communicating<br>• Observing |

## REFERENCES

Brown, P. L., and S. K. Abell. 2007. Examining the learning cycle. *Science and Children* 44 (5): 58–59.

Bybee, R. W. 2011. Scientific and engineering practices in K–12 classrooms: Understanding *A Framework for K–12 Science Education*. *Science Scope* 35 (4): 6–12.

Bybee, R. W., J. A. Taylor, A. Gardner, P. Van Scotter, J. C. Powell, A. Westbrook, and N. Landis. 2006. *The BSCS 5E instructional model: Origins, effectiveness, and applications.* Colorado Springs, CO: BSCS.

Chessin, D. A., and V. J. Moore. 2004. The 6-E learning model. *Science and Children* 42 (3): 47–49.

Eisenkraft, A. 2003. Expanding the 5E model. *Science Teacher* 70 (6): 56–59.

Karplus, R., and R. G. Fuller. 2002. *A love of discovery: Science education, the second career of Robert Karplus.* New York: Springer.

Martin, R., C. Sexton, K. Wagner, and J. Gerlovich. 1997. *Teaching science for all children.* Needham Heights, MA: Allyn and Bacon.

National Research Council (NRC). 2012. *A framework for K–12 science education: Practices, crosscutting concepts, and core ideas.* Washington, DC: National Academies Press.

Ohio Resource Center. 2012. The learning cycle. *http://ohiorc.org/pm/science/ Sci_LearningCycle.aspx.*

Padilla, M. J. 1990. The science process skills. Research Matters—to the Science Teacher. Paper 9004, National Association for Research in Science Teaching, Reston, VA. *www.narst.org/publications/research/skill.cfm.*

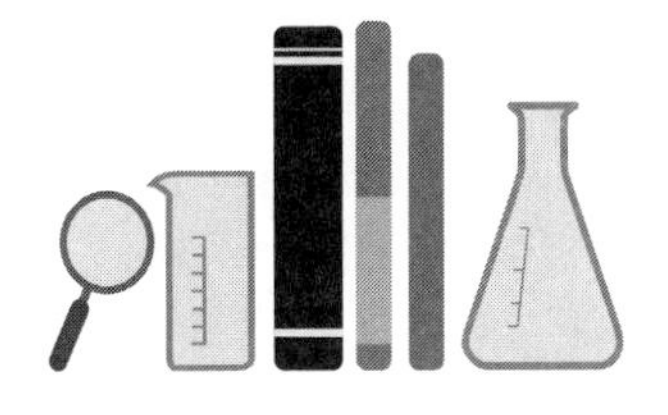

## Chapter 3

# Literacy in the Science Classroom

As I reflect on middle school students and reading, two memories come to mind. Early in my teaching career, I regularly assigned textbook reading to my middle school science students as homework, and then despaired when they didn't seem to know any of the content the next day. Pop quizzes and lectures didn't prove to be effective motivation techniques, and I found myself repeating the same frustrating cycle over and over again. As I've learned more about reading, nonfiction text, and the way students learn, I've come to realize that my students may very well have been doing the required reading. Perhaps they just didn't have the tools to make meaning of the complex text.

Many years later, I was tutoring an eighth-grade student after school that was having difficulty in social studies. We had just started working together, and I was still figuring out how to best help her. While reading a particularly difficult passage in the textbook, I offhandedly remarked, "You know, when I have trouble understanding what I'm reading, I'll stop and re-read that part again. Sometimes even out loud." Shocked, she asked me, "You have trouble understanding what you read?" In that moment, I realized that many students have only seen teachers and adults read fluently and without confusion. The simple admission that, yes, sometimes text is challenging for adults, was a game-changer for her. Students gain so much when we admit our challenges (or even make mistakes in front of them) and share our strategies for overcoming those challenges.

—Jessica

In recent years, the definition of literacy has expanded from simply reading and writing to include speaking, listening, and viewing images and other graphics. We'll consider each of these components, along with vocabulary instruction, and its role in the middle school science classroom.

## READING NONFICTION TEXT

A popular saying in education is that reading instruction in the primary grades is focused on "learning to read," whereas upper elementary and higher is concerned with "reading to learn." Although there is some truth to that statement, it does not accurately reflect the complex nature of reading. As we now know, reading is more than just decoding words on a page—it is making meaning from text (Rosenblatt 2005). When we frame reading in this way, we see that while young students are learning to decode words and sentences, they are also using text to learn about the world around them. Older students, faced with increasingly complex texts, are learning to effectively navigate them even as they glean information from them. In other words, we are always learning to read *and* reading to learn.

This point of view is especially helpful when we consider the types of text that we ask our students to read in middle school science. Many science textbooks are written above grade level (Budiansky 2001; Chall and Conard 1991) and contain text structures that are challenging for students. Although instruction in nonfiction text has increased in the elementary grades in past years, students still arrive in middle school classrooms needing explicit instruction and support. Five common nonfiction text structures are description, sequence, compare-and-contrast, cause-and-effect, and problem and solution (Oczkus 2014).

The author's purpose in writing the text and the specific content included determine the text structure. Each of the five structures described are characterized by unique organization and signal words. Comparing passages of each structure written about the same topic is an excellent way to understand the differences between the structures. Table 3.1 compares passages written about the digestive system.

## SUPPORTING STUDENTS IN READING NONFICTION TEXT

Textbooks, including those used in middle school science, employ all five of these text structures. In fact, it is not uncommon to find several different structures used in a single page of text, and, as previously discussed, they are frequently written above grade level. For those reasons, we recommend the use of trade books, articles, and other sources of text to develop an understanding of concepts introduced through inquiry whenever possible—an approach echoed by experts in the field (Allington 2002). We cannot, however, deny that textbooks play a central role in many middle school science classrooms. We do recommend teaching and using reading comprehension strategies to help students successfully interact with this challenging text. Two such strategies are close reading and teaching text structures.

Close reading is a strategy that promotes a deep understanding of a text. Over the course of multiple readings, students attend to key ideas and supporting details and reflect on the meaning of words and sentences and how the author develops ideas over the text as a whole. Although it has no set instructional sequence, close reading is a multistep process. Students begin by reading the text individually, focusing on the main ideas and details at first. A brief class discussion or think-pair-share allows the teacher to assess student understanding. A second reading focuses on what is known as the *craft and structure* of the text—that is, the choices made by the author regarding vocabulary and text structures. Another discussion follows to check for understanding. A third reading asks students to synthesize information and ideas. Following a third reading, students journal a response to a text-dependent question that requires them to refer to the text for evidence.

Close reading is an excellent strategy for promoting comprehension, but it is not appropriate in all cases. The text used in the process should be brief, and it must be sufficiently

## TABLE 3.1. NONFICTION TEXT STRUCTURES

| Text Structure and Purpose | Signal Words | Sample Text |
|---|---|---|
| **Descriptive text** provides details and characteristics about an object, class, or group. | For example, including, such as, appears to be<br><br>Also uses sensory details and descriptive language | The digestive system includes the mouth, tongue, esophagus, stomach, and large and small intestines. The tongue is covered with tiny bumps known as *papillae.* Papillae help grip food while you are chewing and also contain the taste buds that detect sweet, salty, sour, and bitter tastes. The esophagus is a muscular tube that connects the throat and the stomach. The stomach is a muscular organ located on the upper left side of the abdomen. The small intestine and large intestine are made of highly folded tissues, which maximize absorption of nutrients and water. |
| **Sequential text** (also known as time-order text) provides the steps and order in which a process happens. | First, next, then, finally, last, after | First, food undergoes mechanical and chemical digestion in the mouth through chewing and the action of enzymes in saliva. The food is turned into a soft lump called a bolus. Next, the bolus travels down the esophagus by the action of muscle contractions known as *peristalsis.* The bolus enters the stomach, where acid, enzymes, and muscle contractions break it down into a liquid called *chyme.* After that, the chyme is released into the small and large intestines, where nutrients and water are absorbed into the bloodstream. |
| **Compare-and-contrast text** explores the similarities and differences between objects or classes of objects. | Similar, different, same, alike, unlike, in contrast, on the other hand | The small and large intestines are both long, muscular tubes that help move water and nutrients from the digested food into the body. The small intestine is actually longer than the large intestine, but narrower. Its function is to absorb nutrients from the digested food. In contrast, the large intestine is shorter and very wide. Its function is to absorb water and salts from the undigested food, and to get rid of the waste. |
| **Cause-and-effect text** describes events or phenomena and identifies reasons for why the events or phenomena occur. | Since, because, if … then, as a result, so, leads to, consequently | When a bolus enters the stomach, it must be broken down further before passing to the intestines. The stomach is a muscular organ that secretes acid and enzymes that break down food. The stomach also contracts periodically, churning and mixing the contents. As a result, the bolus is turned into a liquid called *chyme*. The chyme is now ready to pass into the small intestine. |
| **Problem and solution text** introduces a problem and presents one or more solutions. | As a result, because of, leads to, problem, issue, cause, consequently | The gallbladder is a small organ that sits under the liver. It stores bile and releases this into the small intestine to help digest fats. Sometimes, the bile can crystallize and form gallstones, which can cause pain and nausea. If a gallstone blocks the opening to the gallbladder, bile can become trapped, causing the gallbladder to become inflamed. This inflammation can lead to severe pain and fever. Antibiotics or surgery to remove the gallbladder are two ways to treat this inflammation. |

complex for students to read and re-read over multiple days. The reading level of a text is not the sole determinant of a text's complexity. Instead, the text's vocabulary, sentence structure, coherence (how ideas connect to one another), organization, and the amount of background knowledge needed play important roles in determining complexity (Shanahan, Fisher, and Frey 2012). Of course, the students doing the reading are also an important consideration when contemplating text complexity and appropriateness. Students who possess significant amounts of background knowledge will be more successful with a complex text than will those with little or no prior knowledge.

## TEACHING TEXT STRUCTURES

Another effective strategy for promoting comprehension of nonfiction science texts is to explicitly teach the five text structures previously described (Lipson 1996). This instruction does not have to be extensive or long; mini-lessons are an effective way to introduce students to text structures. Consider the following example from a sixth-grade classroom:

Ms. Jones draws students' attention to a passage from a trade book describing a chemical reaction. She explains that this passage is an example of sequential text, or a text that presents the steps of a process in order. She also tells students that signal words such as *first, next, then, last,* and *finally* often indicate a sequential text. As Ms. Jones reads the passage aloud, she underlines the signal words found in the text and records the steps of the process as a numbered list. Next, she distributes copies of a similar passage to students and releases them to read and mark up the text in the same way. She circulates as students work, providing support to them as needed. After students have finished reading, they reconvene for a class discussion of both the signal words and the passage content. Following the mini-lesson, Ms. Jones posts an anchor chart with the name of the text structure, purpose, and signal words for future reference.

This entire instructional sequence (minus independent work by the students) takes no more than 10 minutes and provides powerful, just-in-time support as students navigate nonfiction text. Of course, teachers should revisit this topic periodically to review and provide students opportunities for extended practice.

Buehl (2011) notes that because students often view a nonfiction text as a random collection of facts rather than an organized piece that provides details to explain relationships between ideas, teachers should help make text structures visible by annotating the text with either letters (*C* for cause; *E* for effect) or phrases ("problem," "cause of problem," "solution"). He also recommends using a series of questions that lead the reader to understand the meaning of a piece of writing. Sample questions for four of the text structures discussed earlier are listed in Table 3.2.

### TABLE 3.2. TEXT STRUCTURES AND QUESTIONS

| Text Structure | Sample Questions |
|---|---|
| Description | • What is the concept?<br>• What category of things does this belong to?<br>• What are its important characteristics?<br>• What are examples of it? |
| Compare-and-contrast | • What is being compared and contrasted?<br>• What makes them alike or similar?<br>• What makes them unalike or different?<br>• What are the most important qualities that make them similar (and different)?<br>• In terms of what is most important, are they more alike or more different? |
| Cause-and-effect | • What happens?<br>• What causes it to happen?<br>• Will the result always happen? Why or why not? |
| Problem-solution | • What is the problem?<br>• Who has the problem?<br>• What is causing the problem?<br>• What are the effects of the problem?<br>• Who is trying to solve the problem?<br>• What solutions are recommended or attempted? |

*Source:* Adapted from Buehl 2011.

We recommend teaching each text structure individually to give students time to master and practice each one in turn. Students tend to find cause-and-effect and problem-solution texts more challenging than the other text structures, so we recommend saving those two for last. Focusing on text structures is a wonderful opportunity for cross-disciplinary collaboration, because these same structures are used in the texts students encounter in other subjects such as social studies.

## VOCABULARY INSTRUCTION

Vocabulary demands in science classes are high. Researchers have found that the vocabulary load in secondary science textbooks approaches that of a foreign language course (Groves 1985; Yager 1983). That doesn't even include what researchers call *academic vocabulary*—words that cut across disciplines, are necessary for effective learning, and often appear in assessments (Baumann and Graves 2010; Nagy and Townsend 2012). Words such as *increase, decrease,* and *analyze* are not part of students' daily lexicon, nor are they specific to one content area. Explicitly teaching these words and highlighting their occurrences across the curriculum help students successfully engage in instructional activities. A final, science-specific consideration is that many terms are used in conversational English in ways that are inconsistent with their meaning in science. Consider the words *theory, hypothesis,* and *argument.* All three are essential in a science classroom, but they have vastly different meanings in everyday English.

Explicit instruction around key vocabulary—both content-specific and academic—is thus essential for learning science content and for reading comprehension. Inquiry-based instruction provides teachers with a huge advantage in terms of vocabulary development. To illustrate this advantage, consider how vocabulary was handled in traditional science classrooms. In many cases, students would be assigned to copy definitions from a glossary at the start of a unit. Students who complied would do so mindlessly and without any real understanding of the concepts involved. Resulting assessments would be exercises in memorization and not tests of true understanding.

In an inquiry-based, learning-cycle approach, vocabulary is not front-loaded. Instead, terms are introduced in a "just-in-time" manner, as students develop a need to know as a result of their participation in inquiry activities. Students are also able to actively participate in the creation of definitions because they have had concrete experiences with the concepts (and, in most cases, nonfiction texts). When learning activities are sequenced so that experiential learning and inquiry come before any expectation of mastery of terms, the resulting definitions are much richer and more likely to stick with students over the long haul. Research tells us that we need a high number of encounters with a word (anywhere from 7 to 12) before we are able to take ownership of it (Stahl and Fairbanks 1989). When students have had opportunities to investigate, talk, and write about concepts before formally learning a term and definition, the repetition supports the transfer of the word and concept to long-term memory.

Teachers can also use strategies that assist students in developing a robust understanding of terms, instead of simple definitions. One research-based method is the Frayer Model (Frayer, Frederick, and Klausmeier 1969) in which students use a graphic organizer to record the definition of a word and its characteristics, examples, and non-examples, as illustrated in Figure 3.1. In Chapter 9, students use a Frayer model to develop an understanding of a chemical reaction. A second strategy, word sorts, involves a collection of words written on index cards. Students arrange the words into categories, providing justification for their choices, and revisit and rearrange categories as they deepen their knowledge (Bintz 2011).

Of course, it is not realistic to expect that students complete Frayer models for every word introduced in a unit or chapter, nor can teachers explicitly teach every single word that students encounter. We recommend paring down traditional vocabulary lists to include only terms that are really necessary for the content students are being asked to master.

## WRITING

Students write for many reasons in science: to document observations, to record procedures and data, and to communicate findings. Both informal (writing in a notebook for personal use) and formal writing (writing for an audience) should be regularly included in science classes. One form of formal writing that has been the focus of much attention recently is an argument.

**FIGURE 3.1. THE FRAYER MODEL: DEVELOPING A DEEP UNDERSTANDING OF VOCABULARY**

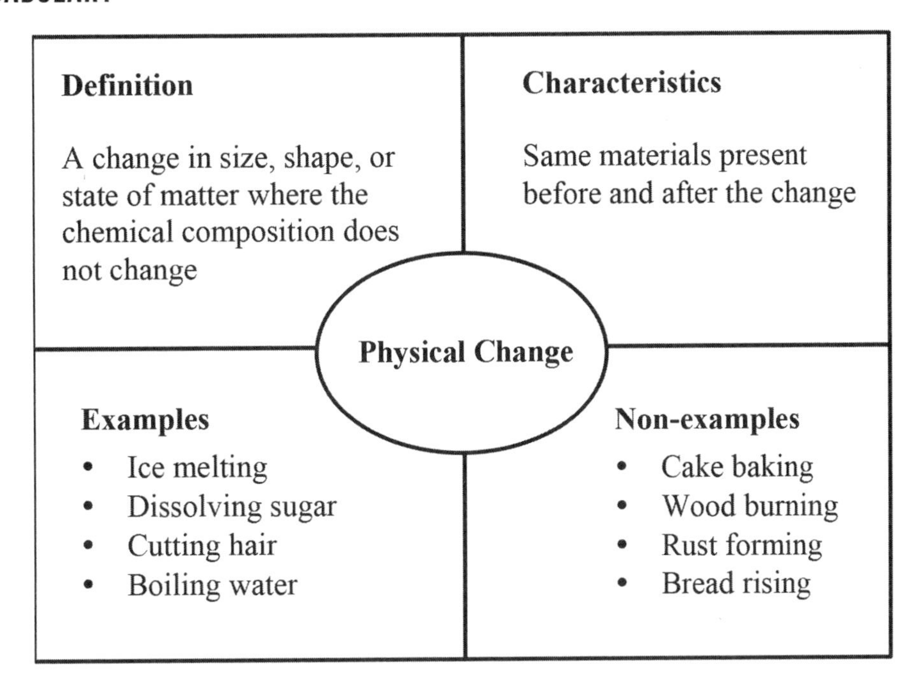


A scientific argument is a piece of writing that includes a claim that is supported by evidence, reasoning that explains why the evidence supports the claim, and a rebuttal that considers and critiques an alternative claim. As with other types of formal writing, students need explicit instruction, modeling, and guided practice as they become proficient in writing arguments. Katherine McNeil's and Joseph Krajcik's *Supporting Grade 5–8 Students in Constructing Explanations in Science: The Claim, Evidence, and Reasoning Framework for Talk and Writing* (2011) provides an excellent overview and tools for successfully implementing argumentation in the middle school classroom. Several of our inquiry-based units, including Chapters 12 and 13, use argumentation during the Explain phase.

Of course, students can (and should!) write in other forms besides argument to share their understanding. From quick writes to formal lab reports, science has a place for all kinds of writing. Teachers can consider the purpose and audience when selecting a format. Is the purpose to consolidate findings and make sense of data? A quick write in a science notebook may suffice. Does the study lend itself to writing to an authentic audience? A letter may be the best choice. In the end, it all comes down to content, context, and purpose.

One effective strategy when planning writing assignments is a RAFT prompt. RAFT is an acronym that stands for four elements (role of the writer, audience, format, and topic) that are specified within the prompt itself. In Chapter 7, "The Genetic Game of Life," students demonstrate their understanding of inheritable traits by writing to the following RAFT prompt:

> You are a **science blogger** for a student news organization, What Do You Know? (WDYK). Your editor has asked you to write a **blog post** telling the story of a couple, Fatu and Shakashare, and **their family history of the disease sickle cell anemia.** Your editor has given you some notes from an interview with the couple for use in your blog post. Your readers are mostly **middle and high school students,** so your editor would like you to also include **an explanation of how the disease is inherited.**

Notice that the four elements of the RAFT prompt are in boldfaced type. This helps teachers check that all needed elements are included and draws students' attention to these crucial components.

Teachers can also use the five text structures (description, sequential, compare-and-contrast, cause-and-effect, and problem-solution) in writing assignments as well as in reading mini-lessons. As students become proficient in recognizing these text structures in nonfiction text, they can also use the texts as examples for their own writing. Linking reading and writing instruction in this way provides powerful support in terms of both reading comprehension and writing ability (Lipson 1996). In Chapter 15, "The Toes and Teeth of Horses," students work with sequential text and then use sequential text as they create a children's book to explain the evolution of horses.

Finally, it is crucial to remember that writing does not simply refer to words on a page. Instead, science and science education are full of instances where meaning is communicated through visual elements such as graphs, diagrams, and maps. Creating these elements is, in fact, a form of writing that complements writing in sentences and paragraphs. Students do not intuitively know how to communicate their understanding through these elements; they require modeling and guidance to make effective use of them in their work. They also need assistance in learning how to select the best type of text with which to convey their meaning (Moline 2011). In Chapter 10, "Nature's Light Show: It's Magnetic!," students create an infographic to explain how the interaction of solar wind and Earth's magnetic field create the aurora. Before they do so, however, they view and critique published examples of infographics. Focusing student attention on infographics encountered in textbooks, trade books, and websites will help build the requisite familiarity needed to generate their own examples in the future.

## SPEAKING

Whether you call it speaking, dialogue, or discourse—conversation is a crucial, yet underused component of literacy in any classroom. Research has demonstrated that in many science classrooms, greater emphasis is placed on conducting investigations than on discussion, argumentation, and making sense of the experiences (Kim and Song 2006). When conversation does happen, it is typically teacher-directed questioning that usually takes an initiate-respond-evaluate form (Cecil and Pfeifer 2011; Mehan 1979; Roller 1989). In many cases, students engage in this dialogue expecting teachers to provide the correct answers, which undermines the value of the inquiry experience (Furtak 2006). In contrast, teachers can engage their students in what Newton (2002) calls *focused talk,* or dialogue that pushes students toward a deeper understanding of the content. Of course, this kind of conversation does not happen spontaneously. Teachers need to carefully plan instruction to facilitate these productive conversations. Cartier, Smith, Stein, and Ross (2013) outline five key practices for orchestrating productive task-based discussions:

1. Anticipate how students are likely to respond to a task.
2. Monitor what students actually do as they work on the task in pairs or small groups.
3. Select particular students to present their work during whole-class discussions.
4. Sequence the student work that will be presented in a given order.
5. Connect different students' responses and connect student responses to key scientific ideas.

The authors note that this model emphasizes planning for conversations in advance, then using "skillful improvisation" during the conversation itself (Cartier, Smith, Stein, and Ross 2013). Of course, with continued practice, teachers can expect to gain proficiency with this model.

As teachers work to move from convergent teaching practices and initiate-respond-evaluate patterns of discourse to more open-ended and student-centered discussion, they should consider the possibility (or eventuality) of unexpected comments. Research has found that when confronted with unexpected ideas, many teachers simply attempt to redirect in an effort to stay on track (Beghetto 2007, 2013; Clark and Yinger 1977). Over time, these redirects can send the message to students that their ideas are not welcome and inadvertently limit risk-taking and creative thinking. While maintaining the integrity of an instructional sequence is important, it is also crucial that teachers attempt to honor unexpected ideas and divergent thinking as often as possible.

Although teacher-led questioning is an essential instructional strategy, it is just as important to provide students with sufficient time to talk to each other about their wonderings, data, and conclusions. Of course, as during any other instructional activity, procedures and routines for productive talk must be taught and practiced. Explicit teaching of strategies for productive conversation mentors students in the ways in which scientists talk to one another. Scientists communicate in specific and precise ways that often differ from everyday conversation, and when teachers take the time to make this type of conversation visible and explicit, they are creating a form of cognitive apprenticeship (Collins, Brown, and Holum 1991) that helps students enter the discipline in an authentic way.

Strategies for helping students engage in productive talk include providing a purpose for conversations, using sentence starters, and teaching students questioning strategies to ask their peers. Some examples of sentence starters for productive discussions include the following:

- I agree because ...
- I would like to add ...
- What evidence do you have for ... ?
- What made you think that?
- Can you give an example of ... ?
- What do you mean by ... ?
- I disagree because ...

Teachers can use the mini-lesson format to model using sentence starters in conversation. The fishbowl technique, in which a small group of students role plays a discussion while the rest of the class observes, can be quite effective in teaching students to use these starters appropriately. Mini-lessons and other explicit teaching of dialogue strategies can be added to any of the inquiry units presented in Part II of this book to support student conversation.

## LISTENING

Listening comprehension is often an underemphasized skill, yet it is an important component of literacy overall. Active listening is not the same as hearing; active listening requires sustained focus and analysis of the information being taken in. Middle school students are often asked to listen to read-alouds (or peers reading aloud), teachers presenting information, and video clips. In many circumstances, setting a purpose before listening, providing a graphic organizer for notes, or doing both will greatly enhance listening comprehension. This purpose can range from listening for specific facts to listening for themes or big ideas, depending on the context of the assignment. For an example of listening comprehension at work during inquiry, see Chapter 7, "The Genetic Game of Life."

## VIEWING

Students encounter a dizzying array of visual information in science classes. Textbooks and nonfiction trade books contain a variety of diagrams, graphs, images, and sidebars that provide important information. However, these elements often confuse students. They might skip graphical elements or pay too much attention to them at the expense of reading the rest of the text (Budiansky 2001). Moline (2011), who calls these elements *visual texts,* notes that because these elements use conventions that differ from traditional texts, students need explicit instruction in how to interpret them. Additionally, because textbooks frequently do not present the same information in written and visual form, students must be able to move back and forth between these types of texts and synthesize information into a cohesive understanding (Buehl 2011). This task presents a high cognitive load for students and requires modeling and practice. As students prepare to create their own infographics in Chapter 10, "Nature's Light Show: It's Magnetic!," they engage in reflective and purposeful conversation around a number of published infographics. This type of instructional activity supports student comprehension of visual texts and allows students to use the visual texts as mentor texts for their own work (Dorfman and Cappelli 2009).

## CONCLUSION

A common thread among all these components of literacy is the need for explicit instruction, even in a science classroom. Even if instruction in these areas is provided in English language arts class, students will need reminders and support to promote transfer of these abilities and skills. However, this explicit instruction does not need to come at the expense of good inquiry science. Teachers can insert 5–10 minute mini-lessons into various points of the learning cycle, in the same way they incorporate just-in-time instruction on how to use a piece of equipment or create a graph. The inquiry units in Part II illustrate how this integration can realistically be achieved in middle school science classrooms.

## REFERENCES

Allington, R. 2002. You can't learn much from books you can't read. *Educational Leadership* 60 (3): 16–19.

Baumann, J. F., and M. F. Graves. 2010. What is academic vocabulary? *Journal of Adolescent and Adult Literacy* 54 (1): 4–12.

Beghetto, R. A. 2007. Ideational code-switching: Walking the talk about supporting student creativity in the classroom. *Roeper Review* 29 (4): 265–270.

Beghetto, R. A. 2013. *Killing ideas softly? The promise and perils of creativity in the classroom.* Charlotte, NC: Information Age Publishing.

Bintz, W. P. 2011. Teaching vocabulary across the curriculum. *Middle School Journal* 42 (4): 44–53.

Budiansky, S. 2001. The trouble with textbooks. *Prism* 10 (6): 24–27.

Buehl, D. 2011. *Developing readers in the academic disciplines.* Newark, DE: International Reading Association.

Cartier, J. L., M. S. Smith, M. K. Stein, and D. K. Ross. 2013. *5 practices for orchestrating productive task-based discussions in science.* Reston, VA: National Council of Teachers of Mathematics.

Cecil, N. L., and J. Pfeifer. 2011. *The art of inquiry: Questioning strategies for K–6 classrooms.* Winnipeg, Canada: Portage and Main Press.

Chall, J. S., and S. S. Conard. 1991. *Should textbooks challenge students?* New York: Teachers College Press.

Clark, C. M., and R. J. Yinger. 1977. Research on teacher thinking. *Curriculum Inquiry* 7 (4): 279–304.

Collins, A., J. S. Brown, and A. Holum. 1991. Cognitive apprenticeship: Making thinking visible. *American Educator* 15 (3): 6–11.

Dorfman, L. R., and R. Cappelli. 2009. *Nonfiction mentor texts: Teaching informational writing through children's literature, K–8*. Portland, ME: Stenhouse.

Frayer, D., W. C. Frederick, and H. J. Klausmeier. 1969. *A schema for testing the level of cognitive mastery.* Madison, WI: Wisconsin Center for Education Research.

Furtak, E. M. 2006. The problem with answers: An exploration of guided science inquiry teaching. *Science Education* 90 (3): 453–467.

Groves, F. H. 1995. Science vocabulary load of selected secondary science textbooks. *School Science and Mathematics* 95 (5): 231–235.

Kim, H., and J. Song. 2006. The features of peer argumentation in middle school students' scientific inquiry. *Research in Science Education* 36 (3): 211–233.

Lipson, M. W. 1996. *Developing skills and strategies in an integrated literature-based reading program.* Boston, MA: Houghton-Mifflin.

McNeil, K. L., and J. S. Krajcik. 2011. *Supporting grade 5–8 students in constructing explanations in science: The Claim, Evidence, and Reasoning Framework for talk and writing.* Boston, MA: Pearson.

Mehan, H. 1979. "What time is it, Denise?": Asking known information questions in classroom discourse. *Theory Into Practice* 18 (4): 285–294.

Moline, S. 2011. *I see what you mean: Visual literacy, K–8.* 2nd ed. Portland, ME: Stenhouse.

Nagy, W., and D. Townsend. 2012. Words as tools: Learning academic vocabulary as language acquisition. *Reading Research Quarterly* 47 (1): 91–108.

Newton, D. P. 2002. *Talking sense in science: Helping children understand through talk.* London: Routledge Falmer.

Oczkus, L. 2014. *Just the facts! Close reading and comprehension of informational text.* Huntington Beach, CA: Shell Education.

Roller, C. M. 1989. Classroom interaction patterns: Reflections of a stratified society. *Language Arts* 66 (5): 492–500.

Rosenblatt, L. M. 2005. *Making meaning with texts: Selected essays.* London: Heinemann Educational Books.

Shanahan, T., D. Fisher, and N. Frey. 2012. The challenge of challenging text. *Educational Leadership* 69 (6): 58–62.

Stahl, S. A., and M. M. Fairbanks. 1986. The effects of vocabulary instruction: A model-based meta-analysis. *Review of Educational Research* 56 (1): 72–110.

Yager, R. E. 1983. The importance of terminology in teaching K–12 science. *Journal of Research in Science Teaching* 20 (6): 577–588.

# Chapter 4
# Integration in Standards-Based Classrooms

Although the natural overlap between science and literacy has been well known for quite some time, standards for both science and English language arts have still been fairly discipline-centric until recent years. The release of *A Framework for K–12 Science Education: Practices, Crosscutting Concepts, and Core Ideas* (*Framework;* NRC 2012), the *Next Generation Science Standards* (*NGSS;* NGSS Lead States 2013), and the *Common Core State Standards, English Language Arts* (*CCSS ELA;* NGAC and CCSSO 2010), reflect a deepened commitment to integration of disciplines. In fact, the *CCSS ELA* include an entire set of standards for literacy in history/social studies, science, and technical subjects. Although this shift may seem to pose yet another challenge for teachers, it actually presents a significant opportunity for authentic, interdisciplinary instruction.

In the introduction to the *Framework*, the authors set forth their goal for K–12 science education:

> The overarching goal of our framework for K–12 science education is to ensure that by the end of 12th grade, all students have some appreciation of the beauty and wonder of science; possess sufficient knowledge of science and engineering to engage in public discussions on related issues; are careful consumers of scientific and technological information related to their everyday lives; are able to continue to learn about science outside school; and have the skills to enter careers of their choice, including (but not limited to) careers in science, engineering, and technology. (NRC 2012)

It is illuminating to consider that three of the five components listed in this goal are directly related to literacy: engaging in discussions, consuming information, and lifelong learning. A fourth, career-related skills, does not explicitly reference literacy but necessarily includes the ability to read, write, and engage in scientific discourse. Clearly, the authors of this document wanted to emphasize the central role that literacy plays in science, as well as other disciplines.

## LITERACY AND THE *NGSS*

### Science and Engineering Practices

It follows, then, that of the eight science and engineering practices defined in the *Framework*, several directly reflect an emphasis on literacy: asking questions and defining problems, constructing explanations and designing solutions, engaging in argument from evidence, and obtaining, evaluating, and communicating information. These practices are, in turn, part and parcel of the *NGSS,* where they appear in the context of disciplinary core ideas. For example, MS-PS1, Matter and Its Interactions, states that students should "gather, read, and synthesize information from multiple appropriate sources and assess the credibility, accuracy, and possible bias of each publication and methods used, and describe how they are supported or not supported by evidence" (NGSS Lead States 2013). This instantiation of the practice of obtaining, evaluating, and communication information closely aligns with several standards from the *CCSS ELA,* as shown in Table 4.1.

### TABLE 4.1. COMPARISON OF ONE READING-RELATED SCIENCE AND ENGINEERING PRACTICE'S ALIGNMENT WITH THE *CCSS ELA*

| *NGSS:* Science and Engineering Practice | *CCSS ELA:* Grades 6–12 Literacy in History/Social Studies, Science, and Technical Subjects |
|---|---|
| MS-PS1-3: Gather, read, and synthesize information from multiple appropriate sources and assess the credibility, accuracy, and possible bias of each publication and methods used, and describe how they are supported or not supported by evidence. | RST.6-8.1: Cite specific textual evidence to support analysis of science and technical texts.<br><br>WHST.6-8.8: Gather relevant information from multiple print and digital sources, using search terms effectively; assess the credibility and accuracy of each source; and quote or paraphrase the data and conclusions of others while avoiding plagiarism and following a standard format for citation. |

*Source:* NGSS Lead States 2013; NGAC and CCSSO 2010.

These companion standards (as well as others that connect to the science and engineering practices throughout the *NGSS*) show that simply reading for content is no longer sufficient. Of course, students continue to use text as an important means to develop the content knowledge that has been nurtured through scientific inquiry, but there is more work to be done. Students are also asked to engage in purposeful, analytical reading—identifying an author's claims, identifying supporting evidence, and considering accuracy and the issue of bias. Students must have a solid understanding of the text to engage in this high level of work, so selecting appropriate texts and providing support for comprehension is essential.

## Crosscutting Concepts

Literacy also plays an important role in developing an understanding of the crosscutting concepts. Crosscutting concepts are overarching lines of thought that span across all science disciplines. They are fundamental to a coherent and cohesive understanding of science and engineering. The *NGSS* (NGSS Lead States 2013) describe the seven crosscutting concepts as follows:

1. *Patterns.* Observed patterns of forms and events guide organization and classification, and they prompt questions about relationships and the factors that influence them.

2. *Cause and effect: Mechanism and explanation.* Events have causes, sometimes simple, sometimes multifaceted. A major activity of science is investigating and explaining causal relationships and the mechanisms by which they are mediated. Such mechanisms can then be tested across given contexts and used to predict and explain events in new contexts.

3. *Scale, proportion, and quantity.* In considering phenomena, it is critical to recognize what is relevant at different measures of size, time, and energy and to recognize how changes in scale, proportion, or quantity affect a system's structure or performance.

4. *Systems and system models.* Defining the system under study—specifying its boundaries and making explicit a model of that system—provides tools for understanding and testing ideas that are applicable throughout science and engineering.

5. *Energy and matter: Flows, cycles, and conservation.* Tracking fluxes of energy and matter into, out of, and within systems helps one understand the systems' possibilities and limitations.

6. *Structure and function.* The way in which an object or living thing is shaped and its substructure determine many of its properties and functions.

7. *Stability and change.* For natural and built systems alike, conditions of stability and determinants of rates of change or evolution of a system are critical elements of study.

Connecting knowledge from one science discipline to another under the umbrella of crosscutting concepts requires students to recognize that there are themes that are common to the science disciplines. Students are not likely to make these connections independently. Intentional and explicit instruction is required. This intentional and explicit instruction can be supported by developing multiple representations of the crosscutting concepts in each science discipline. Multiple representations can take the form of diagrams, formulas, written explanations, flowcharts, maps, drawings, and so on. When students are developing these representations, they are writing; when they are analyzing them, they are reading. As these representations accumulate within a student's body of work, they can begin to be compared across science disciplines.

Consider, for example, the crosscutting concept of stability and change as applied to the units "The Genetic Game of Life" (Chapter 7), "Getting to Know Geologic Time" (Chapter 14), and "The Toes and Teeth of Horses" (Chapter 15). Across these units, students are asked to represent their learning in sketches, written reports, diagrams or drawings, sequences of events, and comparisons. After completing these three units, students will have learned that the inheritance of traits is predictable and generally stable within a population, that changes in the environment can lead to changes in the traits of populations, and that Earth has changed over time. They will have learned each of these things independent of one another. When viewed collectively, these units provide an opportunity to reinforce the concept of stability and change across life and Earth science. This can be seen in the changing surface of Earth, the changes in biodiversity, and the stability of traits from generation to generation. With teacher guidance, students can review the representations of what they have learned and look for these themes. The instruction could be as explicit as asking students how their work represents stability and change in these life and Earth science concepts.

Multiple representations found in texts and a diverse multimodal set of materials can be used along with student-generated representations. Representations ranging from a simulation of changes in Earth over time and the effects of these changes on biodiversity, to an infographic comparing ancestral forms of modern organisms, to a graphic organizer can be used to provide students with additional evidence of stability and change. Careful planning, exposure to multiple representations of a concept, and teacher guidance are necessary parts of helping students develop an understanding of the crosscutting concepts.

## Disciplinary Core Ideas

Literacy is fundamental to students learning the disciplinary core ideas in science. Disciplinary core ideas are grouped into physical science, Earth and space science, life science, and engineering, technology, and applications of science. According to the *NGSS* (NGSS Lead States 2013), for a science concept to be considered core it must meet at least two of the following four criteria (ideally, a disciplinary core idea meets all four):

- Have **broad importance** across multiple sciences or engineering disciplines or be a **key organizing concept** of a single discipline.
- Provide a **key tool** for understanding or investigating more complex ideas and solving problems.
- Relate to the **interests and life experiences of students** or be connected to **societal or personal concerns** that require scientific or technological knowledge.
- Be **teachable and learnable** over multiple grades at increasing levels of depth and sophistication.

# INTEGRATION IN STANDARDS-BASED CLASSROOMS

4

Each unit in this book illustrates the role of literacy in learning the disciplinary core ideas. From defining science terms to developing a pedigree, literacy skills facilitate a student's ability to communicate what she or he knows and can do in science.

## READING NONFICTION TEXT WITH THE *CCSS ELA*

As previously mentioned, the *CCSS ELA* include 10 standards for reading in grade 6–12 history/social studies, science, and technical subjects:

1. Cite specific textual evidence to support analysis of science and technical texts.
2. Determine the central ideas or conclusions of a text; provide an accurate summary of the text distinct from prior knowledge or opinions.
3. Follow precisely a multistep procedure when carrying out experiments, taking measurements, or performing technical tasks.
4. Determine the meaning of symbols, key terms, and other domain-specific words and phrases as they are used in a specific scientific or technical context relevant to grades 6–8 texts and topics.
5. Analyze the structure an author uses to organize a text, including how the major sections contribute to the whole and to an understanding of the topic.
6. Analyze the author's purpose in providing an explanation, describing a procedure, or discussing an experiment in a text.
7. Integrate quantitative or technical information expressed in words in a text with a version of that information expressed visually (e.g., in a flowchart, diagram, model, graph, or table).
8. Distinguish among facts, reasoned judgment based on research findings, and speculation in a text.
9. Compare and contrast the information gained from experiments, simulations, video, or multimedia sources with that gained from reading a text on the same topic.
10. By the end of grade 8, read and comprehend science/technical texts in the grades 6–8 text complexity band independently and proficiently.

### Unpacking the Standards

Even with the close alignment between the science and engineering practices and the *CCSS ELA*, teachers may feel unfamiliar and uncomfortable with these expectations. In this section, we'll briefly unpack each reading standard.

Standards 1 and 2 require students to summarize the text and provide evidence. Science text, as with high-level, complex texts in other disciplines, conveys a lot of information very quickly (Calkins, Ehrenworth, and Lehman 2012). Students need to determine significant ideas to help them sort and sift through details and minutia found in many texts. Notice that standard 2 refers to the "central ideas" of the text, rather than one main idea. These standards reflect the fact that many nonfiction texts are about more than one main idea, thus shifting the language. Once students can answer the question, "What is this text about?" they can go back and locate evidence to support these central ideas. In Chapter 10, "Nature's Light Show: It's Magnetic!," students identify the central ideas of a text about Earth's magnetic field and use these ideas as well as supporting details to summarize the article.

A second notable part of standard 2 is that students need to be able to craft a summary that solely relies on information gleaned from the text in question. Instead of emphasizing prior knowledge and making connections to one's life and experiences, the *CCSS ELA* focus on textual analysis. Although we know that prior knowledge, or schema, plays an important role in student learning at all times, it is worthwhile to have students learn that, at times, constructed responses should focus on only what is presented in a text. Students will undoubtedly need reminders to stick to the main focus as they write summaries and share evidence to support the significant ideas in a text.

Standard 3 states that students should be able to accurately follow multistep procedures, a familiar task for science educators used to facilitating investigations and laboratory exercises. However, it may be novel and informative to consider that our students may need support in comprehending this type of writing. Teachers may benefit from observing students in the lab, specifically watching to determine if they can, in fact, follow the investigation procedures and adjusting instruction accordingly. Many times, our assessment focuses on the students' interpretation of data and their ability to write a conclusion at the end of a lab. Standard 3 helps us remember that reading and comprehending procedures must be given attention as well.

Standard 4 involves vocabulary—determining the meaning of symbols, key terms, and other important words. As discussed in Chapter 3, structuring instruction and designing learning activities that allow students to develop meaningful definitions of content-specific terms is essential. The units in Part II all reflect this just-in-time, student-centered approach to vocabulary development. Sometimes, of course, direct instruction in the meaning of a word or words is needed. This may especially be the case with academic vocabulary. However, students should be responsible for generating definitions for most of the terms deemed essential for any given unit of study.

Standard 5 requires students to analyze a text's structure and explain how parts contribute to the whole. Of course, this necessitates that students can identify the central ideas of the text (see standard 2), but it also lends itself to identifying common nonfiction text structures, including description, sequential compare and contrast, cause and effect, and problem-solution. These text structures, as well as suggestions for instruction, can be found in Chapter 3.

Standard 6 asks students to consider the author's purpose in writing the text. Is the author trying to inform? Persuade? Why did he or she select particular details, anecdotes, or examples to support the main ideas? What did the author choose to include, and what did he or she omit? When students think deeply and critically about the author's purpose and choices in writing a text, they are well on their way to understanding bias—something called for in both the *CCSS ELA* and the *NGSS*.

Standard 7 involves the graphic, or nonlinguistic elements of text. Students who demonstrate proficiency with this standard synthesize information presented in diagrams, illustrations, graphs, tables and figures with that found in text. As discussed in Chapter 3, understanding how to "read" these types of visual elements is hugely important for comprehension of science texts, so it is important to plan structured and scaffolded instruction around these elements.

Standard 8 states that students should be able to differentiate between fact, reasoned judgement, and speculation in a text. Directing students' attention to the evidence presented for claims made in text is one way to assist students in doing so. Science educators might also consider this an opportunity to explore the differing meanings of the word *theory* in science and in conversational English.

Standard 9 asks students to compare findings from investigations and simulations to what they read. The learning-cycle model, discussed in Chapter 2 and implemented in the units in Part II, presents an effective method of doing so. We recommend a "do first, read second" approach that builds an understanding through inquiry before asking students to read and comprehend nonfiction text. Teachers can support this standard in their current lessons and units by asking students, "How do your findings compare to what you read?" and "Is your data consistent with what is presented in the text?"

Standard 10 is the end-goal for all instruction: Students are able to read nonfiction text independently and proficiently. When instruction supports standards 1–9, students are much more likely to achieve standard 10.

## Choosing Texts to Support the Standards

Educators must select texts thoughtfully and intentionally to support students as they develop the skills described in these standards. It is important to note that not every text is suited for all of the standards described earlier. For example, a section from a science textbook will likely present opportunities for students to practice identifying and analyzing the use of nonfiction text structures (standard 5), but probably won't be useful in differentiating between fact, judgement, and speculation (standard 8). Therefore, literacy experts advocate for building text sets (collections of nonfiction texts that offer multiple perspectives), using high-quality journals and digital sources, and channeling resources away from textbooks and toward tradebooks in order to best meet the *CCSS ELA* (Calkins, Ehrenworth, and Lehman 2012).

## WRITING AND THE SCIENCE AND ENGINEERING PRACTICES

Just as reading informational text is closely aligned between the *NGSS* and *CCSS ELA*, so is writing informational text. Again, consider a specific instance of the science and engineering practice engaging in argument from evidence and its corollaries in the *CCSS ELA*, as shown in Table 4.2.

### TABLE 4.2. COMPARISON OF ONE WRITING-RELATED SCIENCE AND ENGINEERING PRACTICE'S ALIGNMENT WITH THE *CCSS ELA*

| *NGSS:* Science and Engineering Practice | *CCSS ELA:* Grades 6–12 Literacy in History/Social Studies, Science, and Technical Subjects |
|---|---|
| MS-LS2-4: Construct an oral and written argument supported by empirical evidence and scientific reasoning to support or refute an explanation or a model for a phenomenon or a solution to a problem. | RI.8.8: Delineate and evaluate the argument and specific claims in a text, assessing whether the reasoning is sound and the evidence is relevant and sufficient; recognize when irrelevant evidence is introduced.<br><br>RST.6-8.1: Cite specific textual evidence to support analysis of science and technical texts.<br><br>W.6-8.1: Write arguments to support claims with clear reasons and relevant evidence.<br><br>WHST.6-8.9: Draw evidence from informational texts to support analysis, reflection, and research. |

*Source:* NGSS Lead States 2013; NGAC and CCSSO 2010.

Many parallels to the reading standards are evident: a focus on evidence, argument, and analytical writing.

## WRITING NONFICTION TEXT

The *CCSS* include 10 standards for writing in grade 6–12 history/social studies, science, and technical subjects:

1. Write arguments focused on discipline-specific content.
2. Write informative/explanatory texts, including the narration of historical events, scientific procedures/experiments, or technical processes.
3. There is no companion to Writing Standard 3 (narrative writing) for history/social studies, science, and technical subjects.

4. Produce clear and coherent writing in which the development, organization, and style are appropriate to task, purpose, and audience.
5. With some guidance and support from peers and adults, develop and strengthen writing as needed by planning, revising, editing, rewriting, or trying a new approach, focusing on how well purpose and audience have been addressed.
6. Use technology, including the internet, to produce and publish writing and present the relationships between information and ideas clearly and efficiently.
7. Conduct short research projects to answer a question (including a self-generated question), drawing on several sources and generating additional related, focused questions that allow for multiple avenues of exploration.
8. Gather relevant information from multiple print and digital sources, using search terms effectively; assess the credibility and accuracy of each source; and quote or paraphrase the data and conclusions of others while avoiding plagiarism and following a standard format for citation.
9. Draw evidence from informational texts to support analysis, reflection, and research.
10. Write routinely over extended time frames (time for reflection and revision) and shorter time frames (a single sitting or a day or two) for a range of discipline-specific tasks, purposes, and audiences.

## Unpacking the Standards

These standards may be more familiar to science educators; however, it is still beneficial to consider each in turn.

Standard 1 focuses on argument writing, a popular genre in science education today. The standard itself is broken into several components that include introducing the claim, supporting the claim with evidence, using words and phrases to connect the claim and evidence (also known as *reasoning* to science educators), and maintaining a formal tone. The resources for argumentation described in Chapter 3 meet this standard beautifully; Chapter 12, "Landfill Recovery," and Chapter 13, "Sunlight and the Seasons," demonstrate how argumentation fits into an inquiry-based, learning cycle unit.

Standard 2 focuses on the production of informative or explanatory text to convey ideas, concepts, and information. This standard, too, is divided into subsections: introducing and developing a topic, using transitions and precise vocabulary, maintaining a formal tone, and including a conclusion statement or section.

Standard 4 involves considering three elements before and during writing: task, purpose, and audience. A paragraph written in a science notebook at the end of an investigation will necessarily differ from a formal argument. A persuasive letter written to the principal requesting recycling measures to be implemented at the school will differ from a lab report. Explicitly stating task, purpose, and audience at the onset of a writing assignment and engaging students in the differences between various types of writing found in science will support students as they work toward this goal. Or, if students are writing independently, asking them to reflect on these three questions will provide support:

1. What am I writing about? (task)
2. What is my reason for writing? (purpose)
3. To whom am I writing? (audience)

Standard 5 also involves purpose and audience, but this time as a focus throughout the writing process. Students collaborate with peers and adults to first plan and, later, review and revise their writing with an eye toward how well the purpose has been met and whether the tone is appropriate for the audience. Providing students with explicit feedback in these areas and with guiding questions or a revision checklist for use with peers helps support this standard.

Standard 6 asks students to use various forms of technology in their writing. This could be achieved by using mapping software to brainstorm or outline, writing using word processing software or collaborative software such as Google Docs, and publishing on the web through web 2.0 technologies.

Standards 7 and 8 focus on research and the appropriate evaluation and use of sources. Media literacy and digital literacy are popular topics in education, with good reason. Now, more than ever, students need to understand how to evaluate the quality and accuracy of information found online as well as how to follow usage guidelines for images and other media.

Standard 9 reiterates the importance of evidence, but this time in writing. If students are proficient with this standard, they are able to pull evidence from informational texts and use that evidence in their own writing. This standard aligns closely with standard 1 (argumentation) and standard 2 (informative and explanatory texts).

Standard 10, as in the reading standards, is the end-goal, in which students write routinely and independently over varying time frames and for varying purposes.

## Understanding the Standards' Benefits

Even though teaching students how to write about their learning and engaging in the writing process takes time, they are worthwhile endeavors for science educators. Research demonstrates that students' understanding of concepts and scores (even on multiple choice assessments) improves when they are given opportunities to write about their learning (Reeves 2000). These improved outcomes seem to occur even when students have moved on to learn about a new topic, suggesting that writing about one's learning ultimately develops habits of mind that facilitate future learning. The key is that students are learning, through explicit instruction and scaffolded learning experiences, how to structure writing, select details and evidence, and elaborate on concepts. If students are just assigned writing without the accompanying instruction, they are more likely to simply "move information from one place to another" (Calkins, Ehrenworth, and Lehman 2012).

## IMPLICATIONS FOR SCIENCE TEACHERS

Although this focus on reading and writing may seem new and overwhelming at times, middle school science teachers can take comfort in the fact that they have a great ally in their English language arts colleagues. The *CCSS ELA* for English teachers and history or science teachers are parallel so that the science and technical standards describe discipline-specific skills and examples of the broader standard for English language arts. This means that within a middle school team, students might practice determining the central ideas of texts in both English language arts and science classes. This type of coordination promotes transfer and allows teachers to share resources and use a common language with students. Science teachers can also feel empowered knowing that they can implement fully integrated inquiry units, such as the ones found in Part II, while adhering to current research and national standards.

## REFERENCES

Calkins, L., M. Ehrenworth, and C. Lehman. 2012. *Pathways to the Common Core: Accelerating achievement.* Portsmouth, NH: Heinemann.

National Governors Association Center for Best Practices and Council of Chief State School Officers (NGAC and CCSSO). 2010. *Common core state standards.* Washington, DC: NGAC and CCSSO.

National Research Council (NRC). 2012. *A framework for K–12 science education: Practices, crosscutting concepts, and core ideas.* Washington, DC: National Academies Press.

NGSS Lead States. 2013. *Next Generation Science Standards: For states, by states.* Washington, DC: National Academies Press. *www.nextgenscience.org/next-generation-science-standards.*

Reeves, D. B. 2000. Standards are not enough: Essential transformations for school success. *NASSP Bulletin* 84 (620): 5–19.

## Chapter 5

# Getting Started With the Inquiry Units

Trying new techniques in your classroom is exciting, but also challenging. Here are some things to consider as you prepare to teach the inquiry units. Of course, we recommend that you first read the unit from start to finish. This step is crucial because it helps you understand the flow of instruction and appreciate the integration of inquiry and literacy practices and the way they reflect the standards.

## GATHER THE BOOKS AND MATERIALS

Each unit heavily depends on the nonfiction texts listed at the beginning of the unit. Some of these texts are trade books; others are found online. If you will need trade books, go to the library several weeks before you begin the unit. If the titles are not available at your local library, you may be able to borrow them through an interlibrary loan program. You may also find good used copies at a secondhand bookstore or online through Amazon or Alibris. If you need to substitute titles, please make sure the new books are well suited for their intended purpose in the lesson. You will also want to review the books for scientific accuracy and check the reading level of each text. We have included Flesch-Kincaid levels, published Lexile levels, or both for each text. Flesch-Kincaid levels are expressed as a grade level, so a reading level of 6.5 indicates that a text is suitable for students in sixth grade. In general, Lexile levels of 955–1,155 are an appropriate target for middle school texts. However, not all students arrive in middle school reading at grade level. Consider the needs of your students as you make decisions about texts.

Most materials for the inquiry units are common items or scientific apparatus that you may already have in your classroom. A few materials may be items that you can easily purchase from a science supply company. There are also many online resources, ranging from articles to interactive simulations. In addition to providing the web address, we also provide QR codes for your convenience. These codes allow you to use a scanning application on your smartphone or tablet to quickly access these resources. For videos, you may notice that we are using SafeShare, a free online tool that removes advertisements and other elements from YouTube and Vimeo videos and creates a permanent link that can be used in the classroom. SafeShare links should be accessible on your school's servers. If for some reason they are not, you may be able to download the videos at home and transfer them to your school computer with a USB drive.

## CONSIDER THE SKILLS AND KNOWLEDGE STUDENTS WILL NEED

Each unit assumes that students bring a certain level of knowledge and experience with them to science. We have included a brief summary of these prerequisites in the overview found at the beginning of each chapter. As you preview and read the unit, we recommend that you keep an eye out for science, reading, or math skills that students will need in order to be successful. If you know that students do not have the requisite skills, you will want to teach them those skills before proceeding. In some cases, you might even find a way to teach the skills during the unit, perhaps through a "just-in-time" teacher-led mini-lesson. Sometimes skills are better learned and more meaningful if they are taught in the context of the content being studied.

## CREATE A SUPPORTIVE CLASSROOM ENVIRONMENT

The units in this book involve a great deal of student communication and collaboration. Expect high interest and energy as students conduct the investigations, and be prepared for a great deal of teacher-student, student-teacher, and student-student discussion. Students need to feel comfortable and confident in sharing their thinking, so it is essential that you review (or institute) expectations for respectful, productive classroom dialogue. Modeling discussion strategies and providing sentence starters, as discussed in Chapter 3, are helpful practices. Students who are accustomed to a more traditional science class, where a teacher provides most of the information through lecture and asks students to simply memorize and recite content, may be uncomfortable with this constructivist approach that values process and thinking. Provide support and feedback to encourage these students, but do not give in to the temptation to do the intellectual work for them. They are more than capable and will feel a great sense of empowerment after completing the investigation or entire unit.

## PREPARE TO MANAGE MATERIALS

Given the high energy and levels of discussion that characterize these units, you will need (and want) to be fully engaged with your students as they test, discover, and reason. You will also need to be a careful observer of students at work, because this is an important form of formative assessment in many of the units. As a result, it is extremely helpful to have systems in place so that students can manage the materials, freeing your time to focus on conversing with students and assessing their work. This system might be as simple as creating baskets or trays of materials for each group and designating one student per group to retrieve and return materials.

## PREPARE FOR A SAFE CLASSROOM ENVIRONMENT

Hands-on, active science is effective and exciting for students and teachers alike. It is essential that, just as you prepare to manage materials and scaffold instructional activities, you also address potential safety issues associated with laboratory and field activities.

You can do so by adopting, implementing and enforcing legal safety standards and better professional safety practices in the science classroom and laboratory. Additionally, you should also review and follow your school's local polices and protocols (e.g., chemical hygiene plan, Board of Education safety policies, and so on).

Throughout this book, safety considerations are provided for laboratory investigations. These considerations are based, in part, on use of the recommended materials and instructions, legal safety standards, and better professional practices. Please note that selecting alternative materials or procedures for these activities may jeopardize the level of safety, so modify investigations with caution. Additional standard operating procedures can be found in the National Science Teacher Association's Safety in the Science Classroom, Laboratory, or Field Sites *(www.nsta.org/docs/SafetyInTheScienceClassroomLabAndField.pdf)*. Students should be required to review the document or one similar to it within the context of a classroom discussion of safety practices. Students and parents or guardians should then sign the document acknowledging procedures that must be followed for a safer learning experience in the laboratory.

## PREPARE GUIDING QUESTIONS TO ENCOURAGE THINKING

A testable question is introduced early in each unit. The testable question frames the inquiry. Guiding questions support students as they work to understand the relevant concepts. Most of these questions do not have right or wrong answers. They are designed to foster discussion and open sharing of ideas. Questions of this nature allow you to peek into your students' thinking. Their responses will expose more about what students know than any multiple-choice questions at the end of the chapter ever could. Listen closely to what your students are saying. They are just as likely to show depth of understanding as they are to uncover misconceptions. In each unit, we have provided sample guiding questions. As you get ready to teach the units, prepare additional questions for follow-ups to the ones we have provided or rephrase ours, if necessary. Questions are critical to inquiry in more than one way!

## USE FORMATIVE AND SUMMATIVE ASSESSMENT

As you and your students work through the units, you will find that we focus much more on formative than summative assessment. As do many others, we believe that formative assessment informs instruction and leads to better outcomes. Only formative assessments are used in the Engage and Explore phases, while students are developing understanding. Formative assessments allow you to collect evidence of student learning and to use that evidence to make some instructional decisions. You may find that students need extra support to complete some tasks or that they can work independently earlier than you expected. Either way, it is important to adjust your instruction accordingly. Formative assessments are typically not graded. The purpose of formative assessment is to gather data to inform instructional decisions.

Summative assessments are typically found in the Explain phase of the inquiry units. The purpose of summative assessment is to evaluate student learning. This is typically where a grade is assigned. We have provided two rubrics for your use. The Science and Literacy Rubric found in Appendix 1 assesses content knowledge and literacy skills and is meant to be used with the summative assessments in the Explain phase and occasionally in the Expand phase. In our experience, rubrics are extremely helpful for identifying and describing the desired outcomes. However, they do not always translate well into letter grades. We have adapted Fanning and Schmidt's (2007) achievement grading approach for use with the units. This approach combines a rubric with a method for assigning a letter grade. The rubric and the achievement grading scale can be found in Appendix 1.

Appendix 1 also includes a Science Process Skills Rubric, which is a formative assessment rubric that can be used periodically throughout the academic year. Whereas *A Framework for K–12 Science Education: Practices, Crosscutting Concepts, and Core Ideas* (NRC 2012) and the *Next Generation Science Standards* (NGSS Lead States 2013) place great emphasis on the eight science and engineering practices, the rubric focuses on the process skills, which are important components of these practices. See Chapter 2 for further discussion of practices versus process skills.

## CONSIDER SEQUENCING INSTRUCTION

Planning the scope and sequence of your instruction over the course of a semester or year is just as important as planning high-quality units. The beauty of the learning-cycle framework is its circular nature, meaning that multiple cycles can be linked together to form extended periods of instruction. This can be achieved in several ways. First, one learning cycle's Expand phase can be followed by another Explore phase that connects related ideas. Alternatively, the activity used as the Expand phase for one cycle could double as an Engage phase for a second, related cycle.

We have provided examples of both approaches in Part II. Chapter 6, "Modeling Cells," focuses on the structure and function of major organelles in both plant and animal cells. This learning-cycle unit could be followed by Chapter 7, "The Genetic Game of Life," in which students learn more about DNA and investigate the mechanism of inheritance for genetic diseases and disorders or by Chapter 8, "Seriously … That's Where the Mass of a Tree Comes From?" which delves into an understanding of the chemical reactions that occur in chloroplasts. You will notice that these chapters are grouped together to facilitate this type of sequenced instruction.

Another example of sequenced instruction is found with Chapter 14, "Getting to Know Geologic Time," and Chapter 15, "The Toes and Teeth of Horses." In the unit on geologic time, students investigate how the fossil record provides a means for understanding, organizing, and describing Earth's history. The activity used in the Expand phase of this unit, comparing extinct and modern-day animals, is identical to the Engage activity in the subsequent unit on evolution. You may wish to meet multiple content standards by teaching these units in succession.

Our experience with this type of instruction is that it transforms science classrooms into hubs of activity and generates enthusiasm among teachers and students alike. We hope it has the same effect in your classroom.

## REFERENCES

Fanning, M., and B. Schmidt. 2007. Viva la revolución: Transforming teaching and assessing student writing through collaborative inquiry. *English Journal* 97 (2): 29–35.

National Research Council (NRC). 2012. *A framework for K–12 science education: Practices, crosscutting concepts, and core ideas.* Washington, DC: National Academies Press.

NGSS Lead States. 2013. *Next Generation Science Standards: For states, by states.* Washington, DC: National Academies Press. *www.nextgenscience.org/next-generation-science-standards.*

# Part II

# Chapter 6
# Modeling Cells

## OVERVIEW

This unit develops an understanding of structure and function in animal and plant cells. Although only selected organelles (nucleus, cell membrane, cell wall, mitochondria, and chloroplasts) are studied, students think deeply about them. This unit uses hands-on investigation and nonfiction text to answer the question, *What role do organelles play in a cell's survival?* Additionally, students reflect on the advantages and limitations of models in conveying structure and function in cells. This unit deals specifically with *eukaryotic cells,* or cells that contain a membrane-bound nucleus and organelles. This unit is an excellent precursor to studying either heredity in "The Genetic Game of Life" (Chapter 7) or photosynthesis in "Seriously ... That's Where the Mass of a Tree Comes From?" (Chapter 8).

This unit assumes that students are able to use a compound light microscope. If students are not familiar with a microscope's parts and proper usage protocols, we recommend conducting a lesson or two about those topics before beginning this unit.

## OBJECTIVES

- Explain the function of the nucleus, cell membrane, cell wall, mitochondria, and chloroplasts.
- Relate the structure of each organelle to its function.
- Develop and use a model to explain the functions of organelles within a cell.
- Identify differences between plant and animal cells.
- Synthesize information from images, investigations, and text.

## STANDARDS ALIGNMENT

### *Next Generation Science Standards* (NGSS Lead States 2013)

#### LS1.A: STRUCTURE AND FUNCTION

- All living things are made up of cells, which is the smallest unit that can be said to be alive. An organism may consist of one single cell (unicellular) or many different numbers and types of cells (multicellular). (MS-LS1-1)
- Within cells, special structures are responsible for particular functions, and the cell membrane forms the boundary that controls what enters and leaves the cell. (MS-LS1-2)

### ***Common Core State Standards, English Language Arts*** **(NGAC and CCSSO 2010)**

GRADES 6–12 LITERACY IN HISTORY/SOCIAL STUDIES, SCIENCE, AND TECHNICAL SUBJECTS

- CCSS.ELA-LITERACY.RST.6-8.1: Cite specific textual evidence to support analysis of science and technical texts.
- CCSS.ELA-LITERACY.RST.6-8.2: Determine the central ideas or conclusions of a text; provide an accurate summary of the text distinct from prior knowledge or opinions.
- CCSS.ELA-LITERACY.RST.6-8.3: Follow precisely a multistep procedure when carrying out experiments, taking measurements, or performing technical tasks.
- CCSS.ELA-LITERACY.RST.6-8.7: Integrate quantitative or technical information expressed in words in a text with a version of that information expressed visually (e.g., in a flowchart, diagram, model, graph, or table).
- CCSS.ELA-LITERACY.RST.6-8.9: Compare and contrast the information gained from experiments, simulations, videos, or multimedia sources with that gained from reading a text on the same topic.
- CCSS.ELA-LITERACY.WHST.6-8.1: Write arguments focused on discipline-specific content.
- CCSS.ELA-LITERACY.WHST.6-8.2: Write informative/explanatory texts, including the narration of historical events, scientific procedures/experiments, or technical processes.
- CCSS.ELA-LITERACY.WHST.6-8.9: Draw evidence from informational texts to support analysis, reflection, and research.

## TIME FRAME

- Twelve 45-minute class periods

## SCIENTIFIC BACKGROUND INFORMATION

Cells are the smallest unit of life. There are two major groups of cells: prokaryotes and eukaryotes. Prokaryotes, which include bacteria and cyanobacteria, lack a distinct nucleus and other organelles. Eukaryotes, which include all other cells, are composed of a membrane-bound nucleus and organelles such as vacuoles, mitochondria, lysosomes, endoplasmic reticulum, and Golgi bodies. Plant cells also contain chloroplasts and are surrounded by a cell wall.

Each organelle serves a specific purpose in the cell's ability to survive:

- The nucleus contains the heredity material of the cell, or DNA. It regulates the cell's activity, including cell division and gene expression.

- The cell membrane serves as the boundary between the cell and the outside world. It is semi-permeable and controls the movement of ions and materials in and out of the cell.
- The mitochondria are the site of cellular respiration in which sugars are broken down and energy is released. This reaction takes place in the folds, or cristae, of the mitochondria. These folds greatly increase the surface area where respiration occurs.
- In plant cells, chloroplasts are the site of photosynthesis during which carbon dioxide and water are converted into sugar using solar energy. Chloroplasts contain stacks (grana) of thykaloids, where the chemical reaction of photosynthesis occurs. The grana greatly increase the surface area available for the reaction to take place.
- In plant cells, the cell wall surrounds the cell membrane and forms the outermost layer of the cell. The cell wall is composed of cellulose, which makes the wall rigid. It provides support and protection for the cell inside. The support provided by cell walls allows plants to maintain their shape, even when they are as tall as a redwood tree.

## MISCONCEPTIONS

A wide variety of misconceptions have been documented about cells, including ideas about their size, shape, and uniformity. For the purposes of this unit, we share some misconceptions about cellular structure and function in Table 6.1.

### TABLE 6.1. COMMON MISCONCEPTIONS ABOUT CELLS

| Common Misconception | Scientifically Accurate Concept |
|---|---|
| • The interior of a cell is completely filled with water.<br>• The interior of a cell is solid.<br>• The interior of a cell is filled with air. | The interior of a (eukaryotic) cell is composed of cytoplasm (a fluid) and membrane-bound organelles. |
| Plants are not made of cells. | All living things, including plants, are made of cells. |

*Source:* AAAS Project 2061 n.d.

## NONFICTION TEXTS

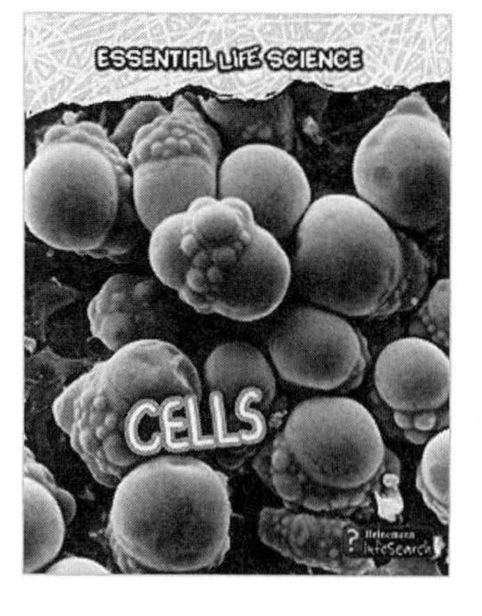

QR CODE 1

*Cells—Infosearch: Essential Life Science* by Richard and Louise Spilsbury (Mankato, MN: Capstone Global Library, 2014); Flesch-Kincaid reading level 6.4, published Lexile level 990L.

This book provides a comprehensive overview of cells, including microorganisms, organelles, and cell division, and it explains how cells form tissue and organs. Beautiful color diagrams and micrographs will draw students in. Using this text in the unit introduces cells as the building blocks of life and differentiates between unicellular and multicellular organisms. The pages of this book that are used in the unit are available through Google Books at *http://bit.ly/1MuHIzP.*

The following nonfiction passages are provided at the end of the unit:

- "What's Inside a Cell?" (Flesch-Kincaid reading level 6.0)
- "What Is the Nucleus?" (Flesch-Kincaid reading level 6.6)
- "What Are Mitochondria?" (Flesch-Kincaid reading level 6.9)
- "What Are Chloroplasts?" (Flesch-Kincaid reading level 6.3)
- "What Is the Cell Membrane?" (Flesch-Kincaid reading level 5.3)
- "What Is the Cell Wall?" (Flesch-Kincaid reading level 4.3)

## MATERIALS

- Compound light microscopes
- Commercially prepared cheek-cell slides
- Microscope slides
- Coverslips
- Pipettes
- Red onion
- Methylene blue stain
- Copies of electron micrographs for each organelle
- Copies of onion root tip cell image
- Dish soap
- Corn syrup
- 4 straws per group of students

- Shallow dish or tray
- Thread
- Effervescent antacid tablets (12 per group of students)
- 100 ml graduated cylinder (1 per group of students)
- Water
- Stop watch (1 per group of students)
- Food coloring
- Quieting solution (laboratory grade)
- *Euglena* culture (living)
- Goggles (sanitized, indirectly vented chemical splash)
- Nonlatex gloves
- Aprons

## SUPPORTING DOCUMENTS

- "Viewing Red Onion Cells" data sheet
- "Viewing Cheek Cells" data sheet
- "Viewing Onion Root Tip Cells" data sheet
- "What's Inside a Cell?" nonfiction text
- "What Is the Nucleus?" nonfiction text
- "Fizzing Tablets" lab sheet
- "What Are Mitochondria?" nonfiction text
- "What Are Chloroplasts?" nonfiction text
- "Modeling the Cell Membrane" activity sheet
- "What Is the Cell Membrane?" nonfiction text
- "What Is the Cell Wall?" nonfiction text
- "Nucleus" graphic organizer
- "Mitochondria" graphic organizer
- "Chloroplasts" graphic organizer
- "Cell Membrane" graphic organizer
- "Cell Wall" graphic organizer
- "My Cell Model" prompts
- "Viewing *Euglena*" data sheet
- "What Are *Euglena*?" prompts

## SAFETY CONSIDERATIONS

- Have students wear goggles, gloves, and aprons during all phases of the investigation, including during setup and cleanup.
- Review safety information in safety data sheets for hazardous chemicals: methylene blue stain (see Appendix 4), effervescent antacid tablets (see Appendix 4 ), and some food colorings.
- Remind students to use caution when working with glassware or plasticware because broken pieces could puncture skin.
- Model safe tweezer use for students.
- Remind students not to eat food or tablets used in investigations.
- Wash hands with soap and water after completing the investigation.

## LEARNING-CYCLE INQUIRY

### Engage

In this phase, students observe plant (red onion) and animal (cheek) cells and discuss their similarities and differences.

1. Introduce the term *cell* and ask students to think-pair-share what they know about cells. (*Note:* For an explanation of the think-pair-share strategy, see page 346 in Appendix 2.)
2. Invite pairs of students to share their ideas with the entire class. Record students' prior knowledge on chart paper or an interactive white board.
3. Explain to students that they will observe two different types of cells and think about their similarities and differences.
4. Conduct the "Red Onion Cells" and "Cheek Cells" activities.

#### RED ONION CELLS

To prepare for this activity, cut a red onion into chunks. Several students will be able to prepare slides from a single chunk.

QR CODE 2

You may choose to show students a video detailing how to prepare a wet mount slide before students complete this activity. One helpful video is available at *http://bit.ly/1LNxQR5.* (*Note:* Stop the video at the 2:20 mark. The rest of the video shows what the students will see when they view their slides. Additionally, the video demonstrates staining with iodine, but students will use methylene blue in this activity.)

Alternatively, you may use the following instructions to demonstrate:

1. Remove an inner layer from an onion section.
2. Use a pair of tweezers to pull off a thin, red outer layer from this section.

3. Spread the layer on a microscope slide.
4. Place one drop of methylene blue stain on top of the onion.
5. Gently lower a cover slip onto the slide, taking care to avoid trapping air bubbles underneath.
6. Allow the slide to sit for 1 minute before viewing.

Students should begin viewing their slides with the lowest power objective and work up to the highest. At each level of magnification, students should draw and describe what they see. You may choose to provide copies of "Viewing Red Onion Cells" for students to use, or you may have students draw and write in their science notebooks.

### CHEEK CELLS

For this activity, students will use commercially prepared slides. Students should begin viewing their slides with the lowest power objective and work up to the highest. At each level of magnification, students should draw and describe what they see. You may choose to provide copies of "Viewing Cheek Cells" for students to use, or you may have students draw and write in their science notebooks.

1. Invite students to discuss their observations of both cells. Use guiding questions to focus student attention:

   - How did the red onion cells and cheek cells compare?
   - What similarities did you observe between the two types of cells?
   - What differences did you observe between the two types of cells?

   During this conversation, introduce the terms *animal cell* and *plant cell.* Explain that the red onion cells are plant cells and the cheek cells are animal cells. Create a Venn diagram on chart paper or an interactive white board and label the three parts of the diagram: (1) animal cells, (2) plant cells, and (3) both plant and animal cells. As students share observations, record their information on the chart. (*Note:* This chart is meant to be a working document. Record all observations and ideas [both correct and incorrect]. There will be opportunities throughout the unit for students to revisit and revise their thinking.)

   Read aloud pages 4–5 of *Cells.* You may choose to project the text using this link: *http://bit.ly/1MuHIzP.*

QR CODE 1

2. Ask students to think-pair-share one piece of information they learned from the text.
3. Invite pairs of students to share their ideas with the entire class. Record new information on the class chart.

QR CODE 3

4. Show the first part of this BrainPop video about cells: *http://bit.ly/1Ini0ac.* (*Note:* Stop the video at the 1:14 mark. The remainder of the video goes on to discuss the functions of each organelle—exactly what students will conclude in this unit.)
5. Draw students' attention to the term *organelles.* Distribute copies of the paragraph "What's Inside a Cell?" Students should read this paragraph individually.
6. Ask students to think-pair-share what they heard about organelles in the video and text.
7. In a whole class discussion, generate a definition for organelles using the information students gained from the video and text. Post the term and definition in a prominent place in the classroom.
8. Introduce the unit question to the students: *What role do organelles play in a cell's survival?*

***Assess this phase:*** This can be an opportune time to assess student proficiency in using a compound light microscope as well as recording observations by drawing. Student observations and ideas about cells and organelles serve as a means of formative assessment. If misconceptions or incorrect information is uncovered, do not correct students at this time. It may be helpful for you to record this information so you can ensure that your instruction in subsequent phases addresses the incorrect concepts.

## Explore

In this phase, students study one organelle at a time through observation of electron micrographs, modeling or inquiry activities, and reading of nonfiction text. Students synthesize what they have learned from these three sources into succinct statements of the function of each organelle.

### PART I: NUCLEUS

1. Introduce the nucleus in name only; do not share any information about it with students. Explain to students that they will participate in several activities to learn about this organelle.

QR CODE 4

2. Project or distribute copies of the electron micrograph of the nucleus found at *http://bit.ly/1OAy9xw.*
3. Ask students to think and write about their observations. Discuss observations as a class.

QR CODE 5

4. Distribute copies of the image of magnified onion root tip cells (available at *http://bit.ly/25bAQme*).

5. Instruct students to cut apart the cells in the image, observe them closely, and sort them into groups. Have students glue the groups of cells onto the "Observing Onion Root Tip Cells" organizer or in their science notebooks. Students should also write a brief justification for their grouping.

6. Discuss the activity with students, using the following guiding questions to encourage critical thinking:

   - What kinds of differences do you notice among the cells?
   - Do you notice any similarities among the cells?
   - How did you decide to group the cells?
   - Did you have any cells that were challenging to group? Why?
   - Why do you think there are these differences among the cells from one root?

7. Distribute copies of the nonfiction text "What Is the Nucleus?" to students. Explain to students that the purpose of reading this text is to learn more about the nucleus and its role in the cell. As students read, they should highlight or underline information that helps them do this. (*Note:* You may need to support students as they read in a variety of ways, including reading in partners or in a small group with teacher support.)

8. Discuss the text as a class, answering student questions and checking for comprehension. Use the following guiding questions to help students connect their observations of the micrographs and onion root tip cells to the information in the text:

   - What information helped you learn about the nucleus and its role?
   - Did you read anything that reminded you of what you saw in the onion root tip cells?
   - Can you make a connection to the images of the nucleus you saw earlier?

9. Distribute copies of the graphic organizer "Nucleus" and ask students to first record what they observed and learned from the activities and the text. After they have done so, explain that their job is to synthesize this information into one or two sentences explaining the purpose of the nucleus. (*Note:* This writing task is an important formative assessment, so it should be completed individually.)

### PART II: MITOCHONDRIA AND CHLOROPLASTS

1. Introduce the next two organelles (mitochondria and chloroplasts) in name, but do not share any information about them. Explain to students that they will participate in similar activities to learn about these organelles.

QR CODE 6

QR CODE 7

2. Project or distribute copies of the electron micrographs of the mitochondria *(http://bit.ly/1WEDfk0)* and chloroplasts *(http://bit.ly/1TIeX2g)* to students.
3. Ask students to think and write about their observations of the organelles. Discuss observations as a class.
4. Explain to students that they will conduct an experiment to learn more about the functions of the mitochondria and chloroplasts. Introduce the experiment by conducting a demonstration: Drop an effervescent antacid tablet into water so that students can observe the fizzing that occurs. Pose the question to students: How do you think breaking a tablet into smaller pieces will affect the reaction?
5. Distribute copies of the "Fizzing Tablets" lab sheet to students and have them conduct the experiment in pairs or small groups.
6. Once students have completed the lab, discuss the results with them. Use the following guiding questions to help students draw conclusions about the relationship between surface area and reaction rates:
   - Which tablet had the greatest surface area? Which tablet had the least surface area?
   - Which tablet reacted the quickest with water? Which tablet reacted the slowest?
   - How does the surface area affect the speed of a reaction?
7. Turn students' attention back to the electron micrographs. Ask students if they can make any connections between their observations and the results of the "Fizzing Tablets" lab. You may need to demonstrate that the lines students observe in the micrographs are folds created by folding a piece of paper accordion style. Ask students the following questions:
   - How might the folds change the surface area of the mitochondria and chloroplasts?
   - Why might mitochondria and chloroplasts need to have a large surface area?
8. Distribute copies of the nonfiction texts "What Are Mitochondria?" and "What Are Chloroplasts?" to students. Explain to students that the purpose of reading this text is to learn more about the organelles and their roles in a cell. As students read, they should highlight or underline information that helps them do this. Continue to support students as they read.
9. Discuss the texts as a class, answering student questions and checking for comprehension. Use the following guiding questions to help students connect their observations of the micrographs and conclusions from the lab to the information in the text:

- What information helped you learn about the mitochondria and its role? The chloroplasts?
- Did you read anything that reminded you of what you saw in the pictures of the mitochondria and chloroplasts?
- Can you make a connection to the experiment you did?
- Why do you think we studied these organelles at the same time?

10. Distribute copies of the graphic organizers "Mitochondria" and "Chloroplasts," and ask students to first record what they observed and learned from the activities and the text. After they have done so, explain that their job is to synthesize this information into one or two sentences explaining the purpose of the two organelles. (*Note:* This writing task is an important formative assessment, so it should be completed individually.)

### PART III: CELL MEMBRANE

1. Introduce the cell membrane in name, but do not share any information about it. Explain to students that they will participate in similar activities to learn about this organelle.
2. Project or distribute copies of the electron micrograph of cell membranes *(http://bit.ly/1sA3v30)* to students.

QR CODE 8

3. Ask students to think and write about their observations of the organelle. Discuss observations as a class.
4. Explain to students that they will work in small groups to investigate the function of the cell membrane. Distribute copies of "Modeling the Cell Membrane" and have them conduct the activity in pairs or small groups.
5. Once students have completed the lab, discuss the results with students. Use the following guiding questions to help them draw conclusions about the characteristics of the material (pores in nylon, no pores in balloon) and the contents of each:

   - What happened to the sand and gravel in the nylon when you placed it in water?
   - What happened to the sand and gravel in the balloon when you placed it in water?
   - What did you notice about the nylon when you viewed it under the microscope?
   - What did you notice about the balloon when you viewed it under the microscope? How did it compare to the nylon?
   - Why do you think you saw the results that you did? How can you explain your results?

6. Distribute copies of the nonfiction text "What Is the Cell Membrane?" to students. Explain to students that the purpose of reading this text is to learn more about the organelle and its role in the cell. As students read, they should highlight or underline information that helps them do this. Continue to support students as they read.
7. Discuss the text as a class, answering student questions and checking for comprehension. Use the following guiding questions to help students connect their observations of the micrographs and conclusions from the lab to the information in the text:
   - What information helped you learn about the cell membrane and its role?
   - Did you read anything that reminded you of what you saw in the pictures of the cell membrane?
   - Can you make a connection to the activity you did?
8. Distribute copies of the graphic organizer "Cell Membrane" and ask students to first record what they observed and learned from the activities and the text. After they have done so, explain that their job is to synthesize this information into one or two sentences explaining the purpose of the two organelles. (*Note:* This is an important formative assessment, so it should be completed individually.)

### PART IV: CELL WALL

1. Introduce the cell wall in name, but do not share any information about it. Explain to students that they will participate in similar activities to learn about this organelle.
2. Project or distribute copies of the electron micrograph of the cell wall *(http://bit.ly/1Vac8MA)* to students.

QR CODE 9

3. Ask students to think and write about their observations of the organelle. Discuss observations as a class.
4. Ask students to return to their drawings of cheek cells and red onion cells from the Engage phase. Students should think-pair-share about the differences between these two types of cells.
5. Invite students to share their observations. Use the following guiding questions to focus student attention on the salient differences:
   - Do you notice any differences among the shapes of the cells?
   - Can you see any parts of the cell that would influence its shape?
   - Why do you think there is a difference in shape between these two types of cells?

6. Distribute copies of the nonfiction text "What Is the Cell Wall?" to students. Explain to students that the purpose of reading this text is to learn more about the organelle and its role in the cell. As students read, they should highlight or underline information that helps them do this. Continue to support students as they read.
7. Discuss the text as a class, answering student questions and checking for comprehension. Use the following guiding questions to help students connect their observations of the micrographs and conclusions from the lab to the information in the text:
   - What information helped you learn about the cell wall and its role?
   - Did you read anything that reminded you of what you saw in the pictures of the cell wall?
   - Can you make a connection to the differences you observed between the cheek and red onion cells?
8. Distribute copies of the graphic organizer "Cell Wall," and ask students to first record what they observed and learned from the activities and the text. After they have done so, explain that their job is to synthesize this information into one or two sentences explaining the purpose of the organelle. (*Note:* This writing task is an important formative assessment, so it should be completed individually.)

*Assess this phase:* This phase is full of opportunities for formative assessment. Monitor student engagement, participation, and questioning during each set of activities, and adjust grouping and expectations accordingly. The highlighted or underlined portions of each text will provide insight into students' ability to extract the most important information from each passage. If you notice that students are highlighting too much or too little, consider pulling them into a small group for a scaffolded reading experience. Student participation in class discussions will serve as another indicator of their comprehension. Finally, the graphic organizer for each organelle should demonstrate understanding of structure and function. If students do not demonstrate this level of understanding, return to more teacher-directed study of that particular organelle before proceeding to the Explain phase.

## Explain

In this phase, students will create two-dimensional and three-dimensional models of cells and use these models to explain the function of each organelle. They will also reflect on the accuracy of the models.

1. Provide students with a variety of cell diagrams from nonfiction texts and online resources. Invite students to carefully examine three to four diagrams and note

the similarities and differences. Ask students to think-pair-share about which diagram they find the most and least helpful and why.

2. Conduct a brief class discussion about the diagrams, inviting students to share their preferences. Be sure to answer any questions that students may have about the diagrams. (*Note:* The diagrams will likely contain additional organelles beyond what students have studied. Explain this to students but tell them to focus only on the organelles studied in the Explore phase.
3. Ask students to each create their own labeled diagram of a cell (either plant or animal). Students should have access to their materials and graphic organizers from the Explore phase, as well as the sample diagrams. Students should submit these diagrams for your review before proceeding to step 4.
4. Students will choose from a variety of materials to create a three-dimensional model of either a plant or an animal cell. Suggested materials include sandwich bags, various shapes of uncooked pasta (rotini, penne, etc.), pipe cleaners, cotton balls, beads, corks, yarn, and paper. In addition to building a physical model, students will also answer the prompts on "My Cell Model" to explain why they chose their materials and how accurate their model is.
5. Provide time for students to share their models and explanations with classmates. This may be done as a gallery walk, in small groups, or as a whole class. (*Note:* For an explanation of the gallery walk strategy, see page 342 in Appendix 2.)

***Assess this phase:*** Student diagrams serve as the final formative assessment for this unit. It is essential that you review these diagrams before allowing students to build their physical models. These diagrams should include all the organelles (cell membrane, mitochondria, nucleus, plus chloroplasts and a cell wall if a plant cell is drawn). Drawings should include relevant details that indicate student understanding. For example, drawings of mitochondria and chloroplasts should include folds. If student diagrams are incomplete or incorrect, have students return to the diagrams and nonfiction text for further reading and review.

Student models and writing serve as summative assessments. Physical models should reasonably represent the cell and organelles studied; however, student writing (the "My Cell Model" handout) will truly reveal the depth of their understanding. For each organelle, students' explanations should indicate an understanding of structure and function. Responses to questions 2 and 3 on "My Cell Model" should indicate reflective thought about the accuracy of their models.

## Expand

In this phase, students will use their models to explain if *Euglena,* ingle-celled photosynthetic organisms, should be considered plants or animals.

1. Explain to students that they are going to observe single-celled organisms called *Euglena* using a light microscope. Use the following instructions to demonstrate how to prepare a wet mount of *Euglena* for observation:
   a. Add a drop of quieting solution to the center of a glass slide.
   b. Add a drop of *Euglena* to the slide.
   c. Add a drop of water and mix using a toothpick.
   d. Gently lower a cover slip onto the slide, taking care to avoid trapping air bubbles underneath.
2. Have students begin viewing their slides with the lowest power objective and work up to the highest. At each level of magnification, students should draw and describe what they see. You may choose to provide copies of "Viewing *Euglena*" for students to use, or you may have students draw and write in their science notebooks.
3. Explain to students that they should use their cell model to determine if *Euglena* are plants or animals. They will write a paragraph explaining which one they think it is and why. You may choose to give students the "Viewing *Euglena*" graphic organizer to help them gather their thoughts before writing.

***Assess this phase:*** Student writing serves as an additional summative assessment. Students' identification of *Euglena* as plants or animals is not the basis for the assessment, and students should not be penalized if they determine that Euglena is a plant. Rather, their writing should be assessed on their understanding of the differences between plant and animal cells and their abilities to use these differences in their classification.

## REFERENCES

American Association for the Advancement of Science (AAAS) Project 2061. n.d. Pilot and field test data collected between 2006 and 2010. *http://assessment.aaas.org/topics/CE#.*

National Governors Association Center for Best Practices and Council of Chief State School Officers (NGAC and CCSSO). 2010. *Common core state standards.* Washington, DC: NGAC and CCSSO.

NGSS Lead States. 2013. *Next Generation Science Standards: For states, by states.* Washington, DC: National Academies Press. *www.nextgenscience.org/next-generation-science-standards.*

Name____________________ Date____________

# Viewing Red Onion Cells

Magnification ____________

What I observed:

______________________________
______________________________
______________________________
______________________________
______________________________
______________________________

Magnification ____________

What I observed:

______________________________
______________________________
______________________________
______________________________
______________________________
______________________________

Magnification ____________

What I observed:

______________________________
______________________________
______________________________
______________________________
______________________________
______________________________

Name_______________________________________ Date____________________

# Viewing Cheek Cells

Magnification ______________

What I observed:

______________________________________________

______________________________________________

______________________________________________

______________________________________________

______________________________________________

______________________________________________

Magnification ______________

What I observed:

______________________________________________

______________________________________________

______________________________________________

______________________________________________

______________________________________________

______________________________________________

Magnification ______________

What I observed:

______________________________________________

______________________________________________

______________________________________________

______________________________________________

______________________________________________

______________________________________________

Name________________________ Date________________

# Viewing Onion Root Tip Cells

Magnification ______________

What I observed:

______________________________
______________________________
______________________________
______________________________
______________________________
______________________________

Magnification ______________

What I observed:

______________________________
______________________________
______________________________
______________________________
______________________________
______________________________

Magnification ______________

What I observed:

______________________________
______________________________
______________________________
______________________________
______________________________
______________________________

Name_______________________________________ Date_____________________

# What's Inside a Cell?

Even though most cells are very small, they are made up of even smaller parts. These parts are called **organelles.** The word *organelle* comes from Latin and means "little tool." You can think of an organelle as a "little organ." It can be difficult to visualize and understand these organelles because they are so small. A good way to understand them is to compare them to the way your body is organized. Your body is made of many organs such as your brain, heart, and lungs. Each organ has a specific job, or role, in keeping you healthy. Some organs work together to perform their roles. For example, your brain, spinal cord, and nerves work together to send messages across your body. We say that organs that work together in your body make up an organ system. Your brain, spinal cord, and nerves make up the nervous system. You depend on all your organs and systems to work correctly to keep you alive and healthy.

Just as your body is made up of organs, cells are made up of organelles. Each organelle plays a specific role in a cell's survival. Some organelles work together, just as organs do. And a cell needs all of its organelles to work properly for the cell to live, grow, and divide. Some of the organelles in cells are the cell membrane, mitochondria, the nucleus, the cell wall, and chloroplasts. All of these organelles have important jobs to do in keeping a cell alive.

Name________________________ Date______________

# What Is the Nucleus?

Cells make up living things such as plants and animals, but they are also alive themselves. Cells are the smallest living thing there is. Sometimes, we call them the "building blocks" of life. Just like other living things, cells obtain and use energy, grow, and reproduce. The organelles that make up each cell have specific roles to play in these activities.

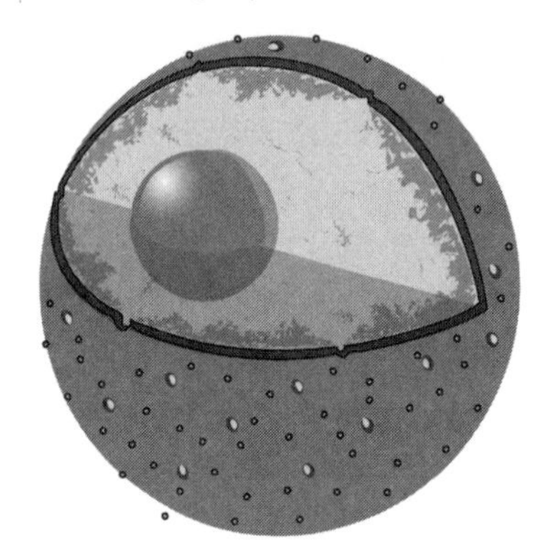

**Nucleus**

One important organelle is the **nucleus.** In the picture of the cell above, the nucleus is the small sphere inside the cell. The nucleus contains the hereditary material of the cell. This hereditary material is what makes each living thing unique. It is also a type of instructional manual or blueprint for many of the cell's actions. One of the important activities that the nucleus directs is cell division.

Cells grow throughout their lives. As they grow, they divide. Most cells divide in a process called **mitosis**. The purpose of mitosis is to create two identical cells. It is the most common way that cells reproduce.

In order to divide, cells go through a series of stages that look like the picture below. What do you notice about the different stages?

Name________________________________ Date________________

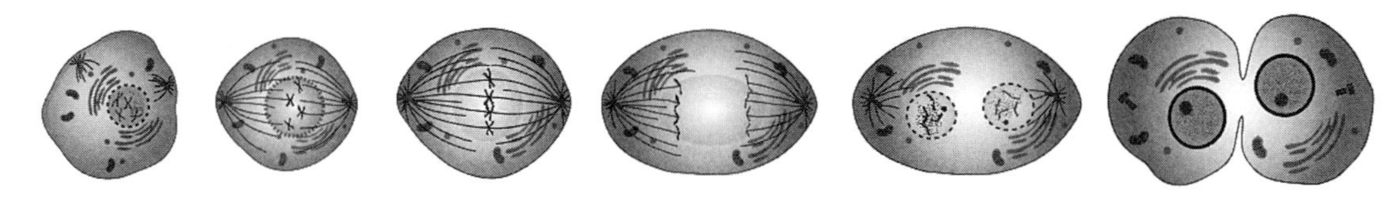

**Mitosis**

Many changes happen inside a cell during mitosis. One change is that the hereditary material is copied. This is important because it means that each new cell has a full copy of the hereditary material. The nucleus makes this and all the other changes of cell division happen. Without the nucleus, important processes such as cell division would not happen.

Name______________________________ Date________________

# Fizzing Tablets

How will breaking an antacid into smaller pieces affect its reaction with water?

Prediction: ____________________________________________

____________________________________________

____________________________________________

## MATERIALS

- 12 effervescent antacid tablets
- 100 mL graduated cylinder
- Water
- Stop watch
- Food coloring
- Goggles
- Non-latex gloves
- Aprons

## SAFETY REMINDERS

- Wear goggles, gloves, and aprons during all phases of the investigation, including during setup and cleanup.
- Follow your teacher's directions regarding safe use of the antacid tablets.
- Be careful working with glass and plastic containers; they can be sharp if broken.
- Immediately wipe up water if spilled on floor.
- Do not eat the antacid tablets.
- Wash hands with soap and water after completing the investigation.

## PROCEDURE

1. Measure 40 ml of water into a graduated cylinder.
2. Add a drop of food coloring to the water.
3. Drop one tablet into the water. Use the stopwatch to time how long it takes the reaction to stop fizzing. Record the time in the data table.
4. Once the reaction stops, empty and rinse the graduated cylinder.
5. Repeat steps 1–4 two more times. Record your results in the data table.
6. Conduct three more trials, but with the tablets broken in half. Record your results in the data table.
7. Calculate the average time for the reaction with the whole tablet and the tablet broken in half.

Name______________________________ Date__________________

## Fizzing Tablets *(continued)*

| Reaction | Trial 1 | Trial 2 | Trial 3 | Average |
|---|---|---|---|---|
| Whole tablet | | | | |
| Tablet broken in half | | | | |

How did breaking an antacid into smaller pieces affect its reaction with water?

______________________________

______________________________

______________________________

______________________________

8. Now use your results to predict how long the reaction will take for a tablet broken into four pieces and for one broken into many small pieces.

   Tablet broken into four pieces: ____________ seconds

   Tablet broken into many small pieces: ____________ seconds

9. Repeat the experiment with tablets broken into four pieces and with tablets broken into many small pieces. Conduct three trials for each. Record your results in the data table. Calculate the average reaction time for each.

| Reaction | Trial 1 | Trial 2 | Trial 3 | Average |
|---|---|---|---|---|
| Tablet broken into four pieces | | | | |
| Tablet broken into many small pieces | | | | |

10. Create a bar graph displaying the reaction time for each tablet.
11. Answer the questions on the next page.

Name______________________________ Date__________________

**Fizzing Tablets** (*continued*)

Breaking the tablets changed the tablets' surface area. Which tablet had the biggest surface area? Which tablet had the smallest surface area?

_______________________________________________

_______________________________________________

_______________________________________________

_______________________________________________

How did changing the surface area of the tablets affect the reaction time?

_______________________________________________

_______________________________________________

_______________________________________________

_______________________________________________

Name________________________________________ Date______________________

# What Are Mitochondria?

As do plants and animals, cells need energy to survive. As it does for other living things, that energy comes from food. Plant cells make sugar using water, carbon dioxide, and the Sun's energy in a process called **photosynthesis.** Animal cells can't create their own food. Instead, they take in food from their environment. Both plant and animal cells need a way to turn that food into a form that they use. Organelles called **mitochondria** do that job for cells.

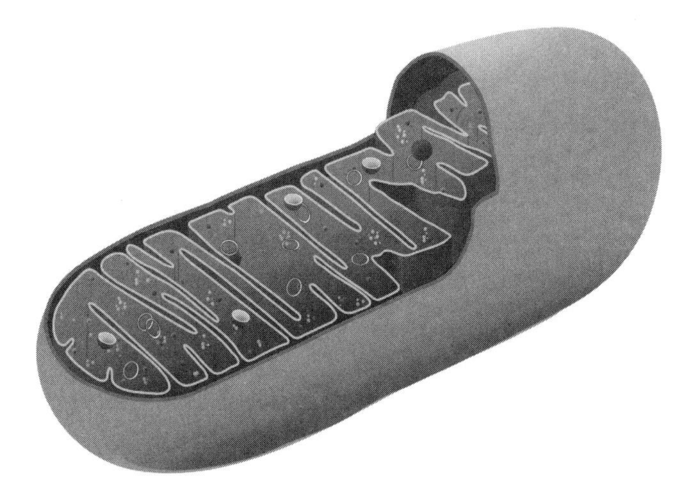

**Mitochondria**

Mitochondria are found throughout plant and animal cells. Inside each one is a series of folds. These folds are where a chemical reaction called **respiration** happens. In the reaction, sugar is broken down and energy is released. The folds inside the mitochondria provide more surface area for this reaction to occur. The more surface area there is, the more energy that can be released.

Mitochondria are extremely important to plant and animal cells. They break down sugar and release the energy that the cells need to carry out all their functions. Without mitochondria, cells would not have the energy they need to grow and reproduce.

Name______________________________ Date__________________

# What Are Chloroplasts?

Cells, as do all living things, need energy to grow and reproduce. And like all living things, cells get their energy from food. Most animals and animal cells take in food from their environment. Plants do not need to take in food. Instead, they make their own food in a process called **photosynthesis.**

In photosynthesis, plants use the energy from the Sun, carbon dioxide, and water to produce a kind of sugar. This sugar is stored by the plant until it is needed. Then, mitochondria in each plant cell break down the sugar and release energy for the plant to use.

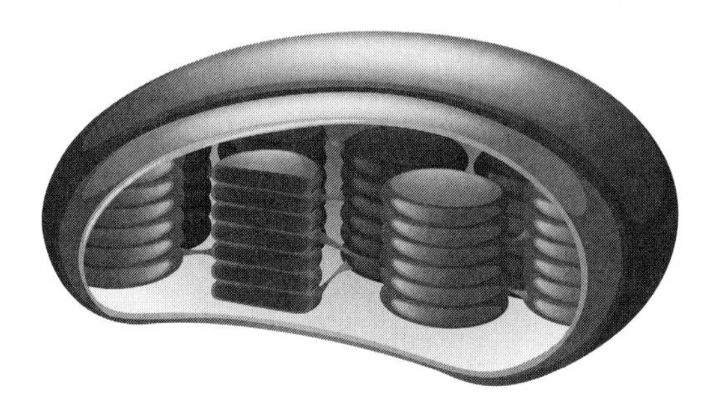

**Chloroplasts**

A special organelle inside plant cells is where photosynthesis takes place. This organelle is called a **chloroplast.** Chloroplasts are green because they contain a green pigment called **chlorophyll.** That pigment is the reason that plants are green in color. Inside each chloroplast are many folds. These folds are where the chemical reaction of photosynthesis takes place. In the chemical reaction, water and carbon dioxide are combined to produce sugar. Energy from the Sun makes this reaction happen.

The folds inside the chloroplasts provide more surface area for this reaction to occur. The more surface area there is, the more sugar that can be created. This energy is needed for the plant cells (and the plant) to grow and reproduce.

Name______________________________ Date__________________

# Modeling the Cell Membrane

Objective: In this activity, you will develop a model of the cell membrane using everyday items.

## MATERIALS

- Knee-high nylon stocking
- Large balloon
- 2 beakers (at least 250 ml)
- Water
- Play sand
- Aquarium gravel
- Measuring cup (¼ cup size)
- Tablespoon
- 2 plates or shallow dishes
- Scissors
- Compound light microscope
- Goggles
- Nonlatex gloves
- Aprons

## SAFETY REMINDERS

- Goggles, gloves, and aprons are to be used during the entire investigation (including during setup and cleanup).
- Be careful working with glass and plastic containers; they can be sharp if broken.
- Always handle scissors carefully.
- Immediately wipe up water if spilled on floor.
- Wash hands with soap and water after completing the investigation.

## PROCEDURE

1. Measure ¼ cup of play sand and 2 tbsp aquarium gravel. Place both materials inside the knee-high nylon.
2. Tie a knot in the nylon to close it.
3. Add ¼ cup play sand and 2 tbsp aquarium gravel to the balloon.
4. Tie a knot in the balloon to close it.
5. Place the nylon and the balloon in separate beakers filled with water. Leave them in the water for a few minutes.
6. Observe both the nylon and the balloon while they are in the water. Sketch and write about your observations.

### Modeling the Cell Membrane *(continued)*

7. Remove both the nylon and the balloon from the beakers. Place each on a plate or a shallow dish.
8. Use scissors to carefully cut open the nylon and the balloon. Observe the materials inside each. How do they compare? Record your observations.
9. Use the scissors to carefully cut a small piece from both the nylon and the balloon. Observe each under the microscope. Sketch and write about your observations.

## OBSERVATIONS

1. In the water

2. Cut open

3. Under the microscope

How can your observations from this activity explain the difference between the contents of the nylon and the balloon after placing both in the water?

______________________________________________________________________

______________________________________________________________________

______________________________________________________________________

______________________________________________________________________

______________________________________________________________________

Name________________________________________ Date____________________

# What Is the Cell Membrane?

Cells are living things. That means that they eat, grow, and reproduce. Inside each cell are organelles that play special roles in the cell's survival. One important organelle is the cell membrane.

The cell membrane is the outer border of animal cells. In plant cells, the cell membrane is not the outermost border. Instead, a **cell wall** surrounds the cell membrane of a plant cell. But the cell membrane plays an important role in both plant and animal cells.

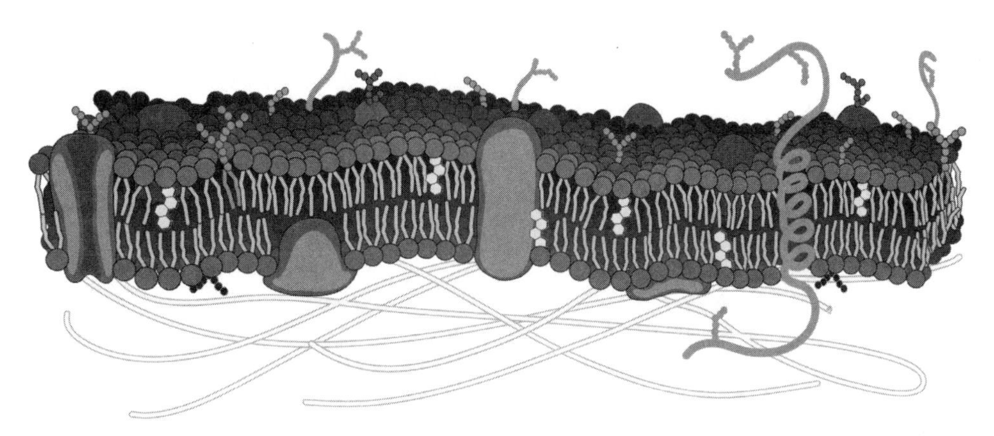

**Cell membrane**

The membrane is made of a double layer of fats, with some proteins mixed in. The membrane is flexible and can change. Tiny holes, or pores, in the membrane allow some materials to move in and out of the cell. These pores are related to the main job of the cell membrane: to control what enters and exits the cell.

For the cell to carry out its functions and survive, it needs to have a carefully controlled interior. Some elements, such as calcium and sodium, need to be present in certain amounts for the cell to survive. The cell membrane keeps some materials in the cell. It lets some materials in and keeps others out. By controlling what goes in and out of the cell, the membrane helps the cell survive.

Name________________________________________ Date______________________

# What Is the Cell Wall?

Plant and animal cells have many things in common. They eat, grow, and reproduce. They both have many of the same organelles that help them survive. For example, both plant and animal cells have a nucleus, mitochondria, and a cell membrane. But plant cells have chloroplasts, and animal cells do not. Another organelle found only in plant cells is the cell wall.

The cell wall is the outermost layer of a plant cell. It surrounds the flexible cell membrane. But the cell wall is not flexible. It is made of a stronger material called *cellulose*. This makes the cell wall hard and rigid. The rigid cell wall is what gives plant cells their box-like shape. It also protects the plant cell inside. The cell wall is just one of the ways in which plant cells are different from animal cells.

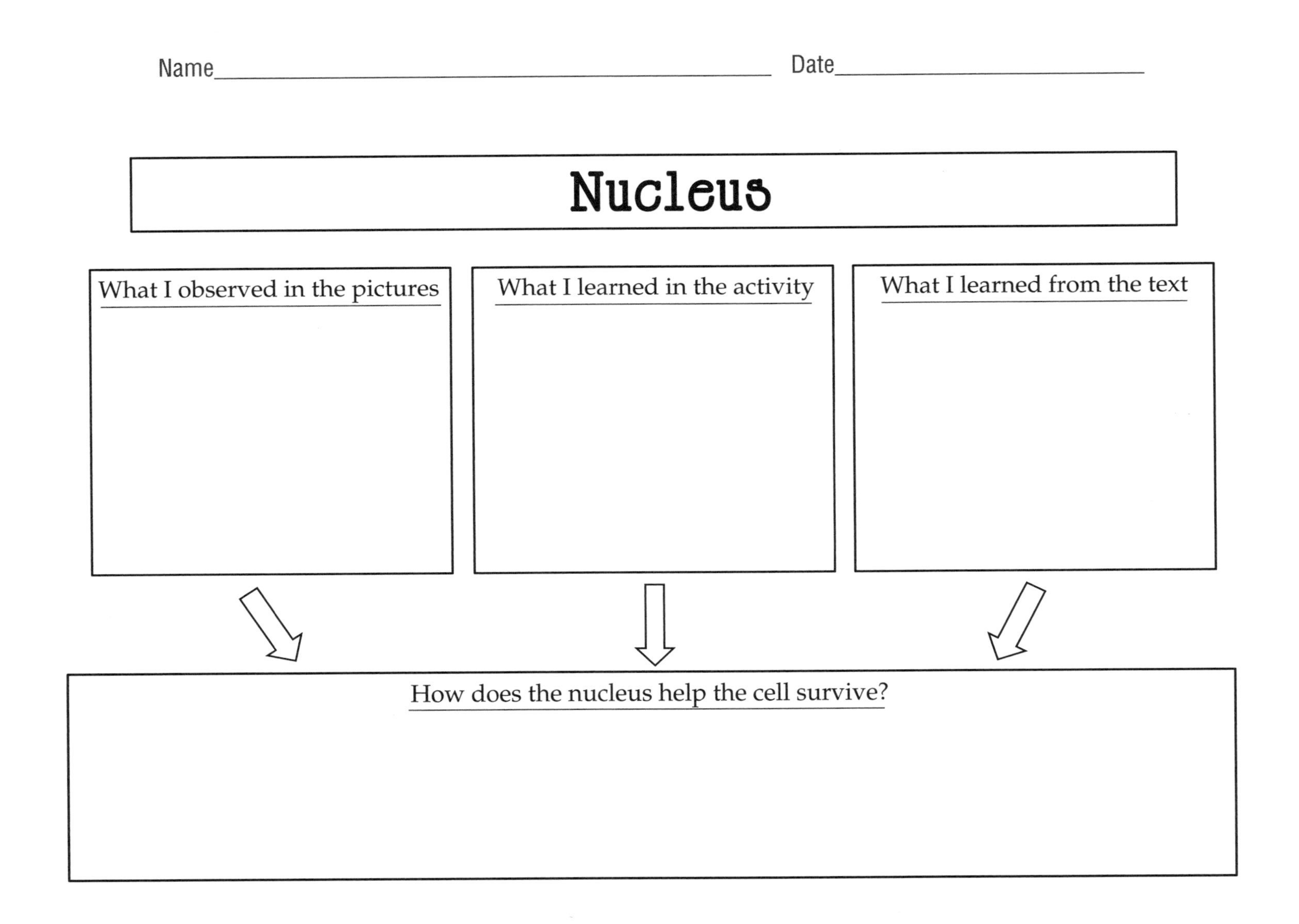
Name
Date
Nucleus
What I observed in the pictures
What I learned in the activity
What I learned from the text
How does the nucleus help the cell survive?

Name____________________ Date________________

# Mitochondria

| What I observed in the pictures | What I learned in the activity | What I learned from the text |
| --- | --- | --- |
| | | |

How do mitochondria help the cell survive?

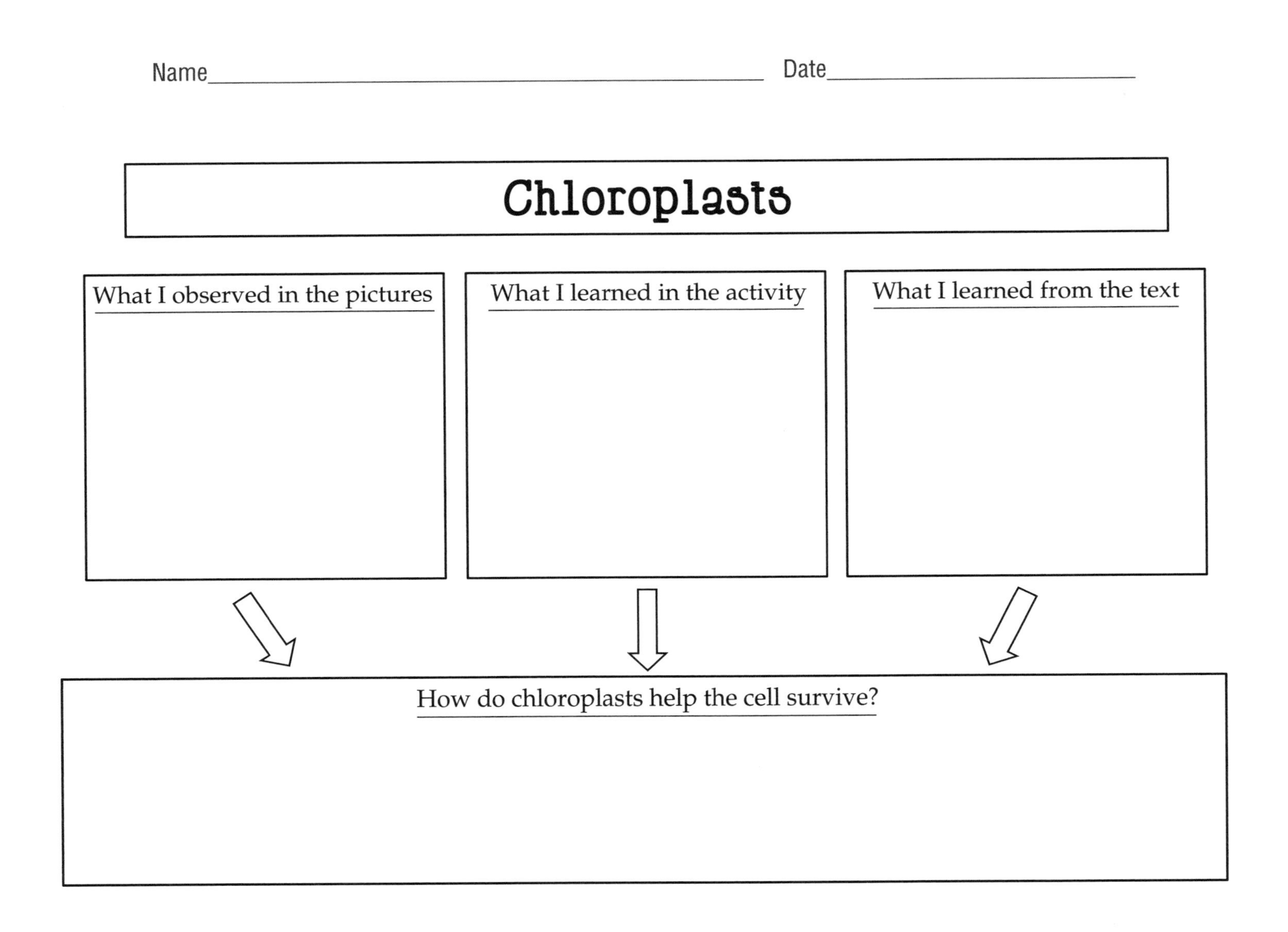
Name
Date
Chloroplasts
What I observed in the pictures
What I learned in the activity
What I learned from the text
How do chloroplasts help the cell survive?

Name________________________ Date________________

# Cell Membrane

What I observed in the pictures

What I learned in the activity

What I learned from the text

How does the cell membrane help the cell survive?

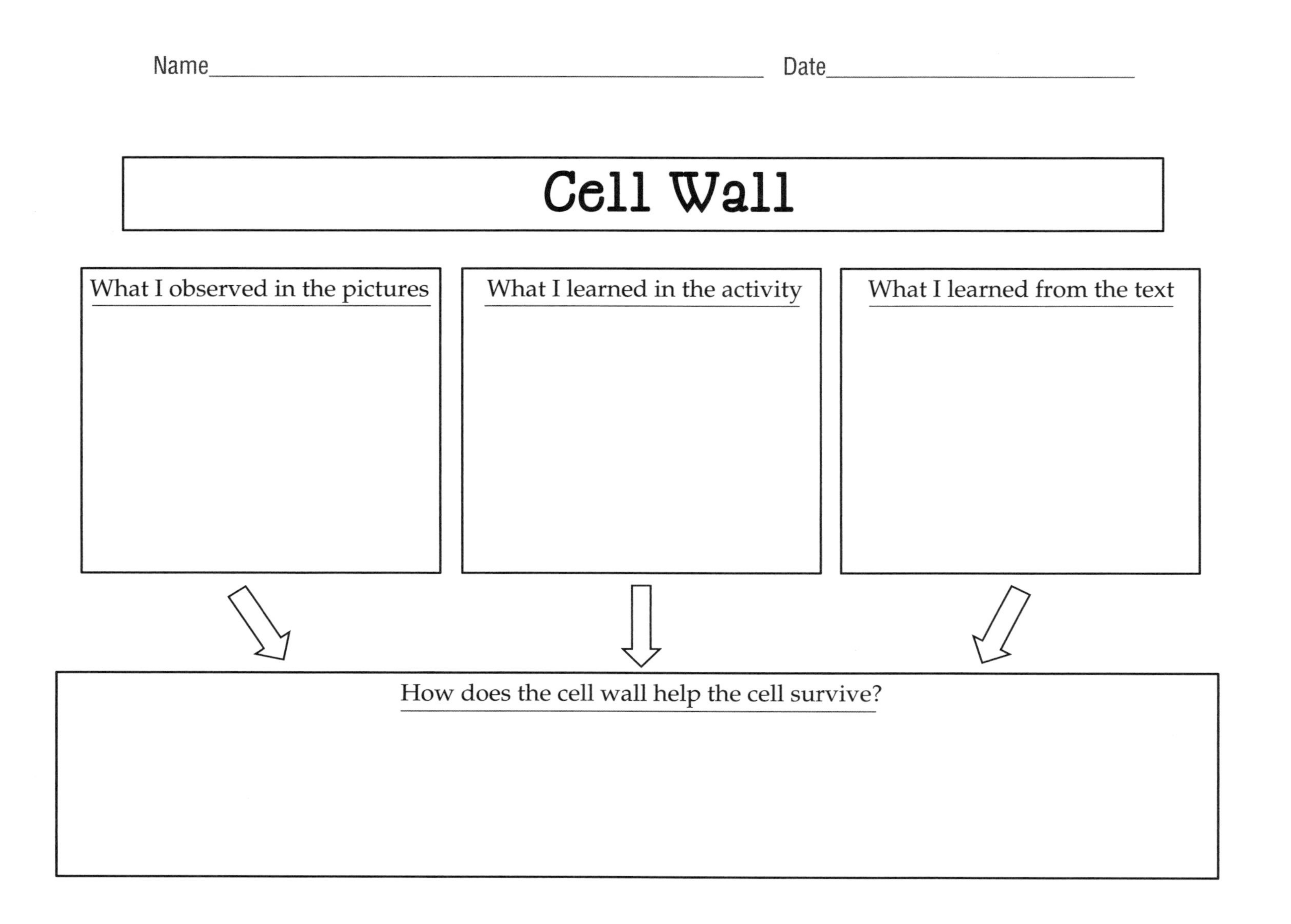
Name
Date
Cell Wall
What I observed in the pictures
What I learned in the activity
What I learned from the text
How does the cell wall help the cell survive?

Name________________________ Date____________

# My Cell Model

Write about each organelle in your cell model. What role does each organelle play in the cell's survival? Why did you choose the materials you did for each organelle?

Name______________________________ Date________________

## My Cell Model (*continued*)

1. In what ways is your model an accurate model of a cell?

_______________________________________________
_______________________________________________
_______________________________________________
_______________________________________________
_______________________________________________
_______________________________________________
_______________________________________________
_______________________________________________
_______________________________________________
_______________________________________________
_______________________________________________
_______________________________________________

2. In what ways is your model not an accurate model of a cell?

_______________________________________________
_______________________________________________
_______________________________________________
_______________________________________________
_______________________________________________
_______________________________________________
_______________________________________________
_______________________________________________
_______________________________________________
_______________________________________________
_______________________________________________
_______________________________________________

Name________________________________ Date________________

# Viewing *Euglena*

Magnification ____________

What I observed:

______________________________
______________________________
______________________________
______________________________
______________________________
______________________________

Magnification ____________

What I observed:

______________________________
______________________________
______________________________
______________________________
______________________________
______________________________

Magnification ____________

What I observed:

______________________________
______________________________
______________________________
______________________________
______________________________
______________________________

Name______________________________ Date______________

# What Are *Euglena*?

Do you think *Euglena* are plants or animals?

______________________________________________

______________________________________________

______________________________________________

______________________________________________

Why do you think this? Use your cell model to help you explain.

1. ______________________________________________

______________________________________________

______________________________________________

2. ______________________________________________

______________________________________________

______________________________________________

3. ______________________________________________

______________________________________________

______________________________________________

**Chapter 7**

# The Genetic Game of Life

## OVERVIEW

This unit develops an understanding of the inheritance of Mendelian traits. Students read about genetic testing, extract deoxyribonucleic acid (DNA) from strawberry cells, simulate monohybrid crosses, learn how to use Punnett squares to predict the frequency of traits in offspring, and construct simple pedigree diagrams. They then investigate sickle cell anemia and other inherited genetic disorders. The unit culminates in a mock medical poster session in which groups of students present their findings. This unit uses hands-on investigation and nonfiction text to answer the question, *How can diseases be inherited?*

This unit assumes that students know and can demonstrate understanding of the nucleus's function in a cell. You might teach Chapter 6, "Modeling Cells," before beginning this unit.

## OBJECTIVES

- Develop and use a model to explain the inheritance of traits.
- Predict the frequency of genotypes and phenotypes in a cross.
- Recognize that each parent contributes equally to the genotype even when the recessive trait is not reflected in the phenotype.
- Represent the inheritance of a trait in a pedigree diagram.
- Explain how a single Mendelian trait is inherited.
- Explain that mutations cause changes in proteins that may be beneficial, harmful, or have no effect on an individual.
- Use research skills to learn about heritable disease.
- Identify the central ideas of a text.

## STANDARDS ALIGNMENT

### *Next Generation Science Standards* (NGSS Lead States 2013)

#### LS1.B: GROWTH AND DEVELOPMENT OF ORGANISMS

- Organisms reproduce, either sexually or asexually, and transfer their genetic information to their offspring. (secondary to MS-LS3-2)

### LS3.A: INHERITANCE OF TRAITS

- Genes are located in the chromosomes of cells, with each chromosome pair containing two variants of each of many distinct genes. Each distinct gene chiefly controls the production of specific proteins, which in turn affects the traits of the individual. Changes (mutations) to genes can result in changes to proteins, which can affect the structures and functions of the organism and thereby change traits. (MS-LS3-1)
- Variations of inherited traits between parent and offspring arise from genetic differences that result from the subset of chromosomes (and therefore genes) inherited. (MS-LS3-2)

### LS3.B: VARIATION OF TRAITS

- In sexually reproducing organisms, each parent contributes half of the genes acquired (at random) by the offspring. Individuals have two of each chromosome and hence two alleles of each gene, one acquired from each parent. These versions may be identical or may differ from each other. (MS-LS3-2)
- In addition to variations that arise from sexual reproduction, genetic information can be altered because of mutations. Though rare, mutations may result in changes to the structure and function of proteins. Some changes are beneficial, others harmful, and some neutral to the organism. (MS-LS3-1)

## *Common Core State Standards, English Language Arts* (NGAC and CCSSO 2010)

### GRADES 6–12 LITERACY IN HISTORY/SOCIAL STUDIES, SCIENCE, AND TECHNICAL SUBJECTS

- CCSS.ELA-LITERACY.RH.6-8.2: Determine the central ideas or information of a primary or secondary source; provide an accurate summary of the source distinct from prior knowledge or opinions.
- CCSS.ELA-LITERACY.RST.6-8.3: Follow precisely a multistep procedure when carrying out experiments, taking measurements, or performing technical tasks.
- CCSS.ELA-LITERACY.RST.6-8.9: Compare and contrast the information gained from experiments, simulations, video, or multimedia sources with that gained from reading a text on the same topic.
- CCSS.ELA-LITERACY.WHST.6-8.2: Write informative/explanatory texts, including the narration of historical events, scientific procedures/ experiments, or technical processes.
- CCSS.ELA-LITERACY.WHST.6-8.6: Use technology, including the internet, to produce and publish writing and present the relationships between information and ideas clearly and efficiently.

- CCSS.ELA-LITERACY.WHST.6-8.7: Conduct short research projects to answer a question (including a self-generated question), drawing on several sources and generating additional related, focused questions that allow for multiple avenues of exploration.
- CCSS.ELA-LITERACY.WHST.6-8.8: Gather relevant information from multiple print and digital sources, using search terms effectively; assess the credibility and accuracy of each source; and quote or paraphrase the data and conclusions of others while avoiding plagiarism and following a standard format for citation.
- CCSS.ELA-LITERACY.WHST.6-8.9: Draw evidence from informational texts to support analysis, reflection, and research.

## TIME FRAME

- Twelve to fifteen 45-minute class periods

## SCIENTIFIC BACKGROUND INFORMATION

Genetics, particularly human genetics, is an exciting yet complex field. This unit deals with what is known as *Mendelian genetics* (classical genetics), or the inheritance of traits that follow the laws proposed by Gregor Mendel. In this unit, we focus on traits that are controlled by one gene and have dominant and recessive versions, or *alleles*. The diseases that students investigate also follow this pattern of inheritance. In high school, students will learn about more complex patterns of inheritance, including multiple alleles and incomplete dominance.

Genes are sections of DNA that code for the production of various proteins. The code is determined by the sequence of nucleotides in that particular section of DNA. Two or more forms, or alleles, of a gene exist, with changes in this sequence of DNA leading to the expression of a different version of a trait.

Sexually reproducing organisms pass DNA to their offspring through their *gametes*, or sex cells. Each parent contributes one-half of the genetic material to the offspring. The combination of alleles, or *genotype*, of an individual determines the *phenotype*, or expressed trait. With Mendelian traits, one allele is said to be dominant to the other, recessive allele. An individual needs to have only one copy of a dominant allele to have the dominant phenotype. Two copies of the recessive allele are needed to express the recessive phenotype.

A few traits in humans, such as freckles, dimples, and a widow's peak, are controlled by one gene. Many other traits, including height, skin color, and eye color, are controlled by the interaction of many genes. Students will likely want to extend their knowledge to all physical traits that they observe in humans, so it is best to familiarize yourself with examples of Mendelian and non-Mendelian traits.

*Mutations* are changes in the sequence of a gene, and they are much more common than many realize. Some mutations are inherited, whereas others are caused by environmental factors such as ultraviolet radiation and viruses. Everyone's DNA contains mutations, and although the word has a negative connotation, few mutations are actually harmful.

Many inherited diseases and disorders are caused by a mutation in a single gene. Of these, most are recessive mutations, meaning that an individual must possess two copies of the mutation to be affected by the disease or disorder. Diseases caused by dominant mutations are rarer because they tend to prevent affected individuals from reproducing before dying.

Sickle cell anemia is an autosomal recessive disease, meaning it is a recessive mutation that is not passed through the X or Y chromosome. (Traits and diseases located on the X or Y chromosomes are known as sex-linked traits.) An individual must have two copies of the recessive allele to have sickle cell anemia. This mutation causes the individual's hemoglobin (a protein found in red blood cells) to become crescent, or sickle shaped. Those irregularly shaped cells can block the passageway of narrow blood vessels, causing pain and inadequate oxygen in parts of the body. The sickle cell mutation is interesting in that individuals who are heterozygous (possessing one normal and one mutant allele) do not suffer from the disease. In fact, being heterozygous for the sickle cell trait has been found to provide protection from malaria, a mosquito-transmitted disease. Scientists hypothesize that the benefit provided by being a carrier of the sickle cell gene explains its high frequency in countries with a high incidence of malaria.

Recent advances in technology have led to the ability to conduct genetic testing to predict an individual's likelihood of contracting certain genetic disorders or to customize treatments for disease. Researchers continue to develop procedures in which gene therapy can be used to correct or treat genetic disorders, although these procedures are not ready for widespread use at this time.

## MISCONCEPTIONS

As with all scientific concepts, students bring incomplete and incorrect understandings of heredity to science class (Radford and Bird-Stewart 1982; Shaw et al. 2008). These misconceptions can profoundly influence the way that a student perceives and receives information during instruction, so it is important for teachers to be aware of possible misconceptions and recognize them in their students' oral and written responses. Table 7.1 lists a few common misconceptions about Mendelian inheritance.

## TABLE 7.1. COMMON MISCONCEPTIONS ABOUT HEREDITY

| Common Misconception | Scientifically Accurate Concept |
|---|---|
| One gene is always responsible for one trait. | Multiple genes often work together to determine a phenotype. Environmental factors also play a role in determining phenotype. |
| One gene with one mutation always causes one disease. | Many inherited diseases and disorders are caused by one gene, but others are caused by the interaction of multiple genes. |
| When an individual has a dominant trait, they must have two copies of the dominant allele. | Dominant traits are expressed when an individual has one or two copies of the dominant allele. |

## NONFICTION TEXTS

QR CODE 1

- "Doctors Get a Jump on Disease With Genetic Testing" by *Minneapolis Star Tribune,* adapted by NewsELA. 2014, *http://bit.ly/1I2NvbC;* Flesch-Kincaid reading level 7.7, published Lexile level 1030L. (QR code 1)

This article from the website NewsELA describes how genetic testing is being used to treat diseases and screen for the possibility of future diseases. NewsELA offers the article in several reading levels.

QR CODE 2

- "Mutations and Disease" by The Tech Museum of Innovation, 2013, *http://bit.ly/1Dkd7zL;* Flesch-Kincaid reading level 10.1. (QR code 2)

This article defines mutation and describes the connection between mutation and disease. Because of its high reading level, we recommend using this article as a read-aloud.

QR CODE 3

- "A Mutation Story" by PBS Learning Media, 2016, *http://bit.ly/210fvXi.* (QR code 3)

This video segment tells the story of a genetic mutation affecting the population of West Africa. The host website, PBS Learning Media, requires the creation of a free account to save the video to your favorites.

QR CODE 4

- "Solving a Genetic Mystery" by Ask a Biologist, Arizona State University School of Life Sciences, n.d., *http://bit.ly/1Vi616e;* Flesch-Kincaid reading level 8.2. (QR code 4)

This article describes Mendel's pea plant crosses and explains how to construct and understand a Punnett square.

QR CODE 5

- "What Is DNA?" (Wonderopolis Wonder of the Day #1329) by National Center for Families Learning, n.d., *http://bit.ly/1R5kn8W;* Flesch-Kincaid reading level 8.1. (QR code 5)

This article provides a basic description of DNA's structure and function. The article includes highlighted terms that provide definitions when scrolled over.

The nonfiction passage provided at the end of the unit is as follows: "Connecting DNA to Heredity" (Flesch-Kincaid reading level 5.7)

## MATERIALS

### Per Student

- Resealable plastic bag
- 2 strawberries (fresh or frozen)
- 2 tsp dish soap
- ½ cup water
- 1 tsp salt
- 2 plastic cups
- 1 coffee filter
- ½ cup cold rubbing alcohol (see safety data sheet in Appendix 4)
- 1 wooden stick (such as a coffee stirrer or popsicle stick)
- 2 dice
- Goggles (sanitized, indirectly vented chemical splash)
- Nonlatex gloves
- Apron

### General Supplies

- Chart paper
- Sticky notes

## SUPPORTING DOCUMENTS

- "Defining DNA" graphic organizer
- "Connecting DNA to Heredity" nonfiction text
- "What's the Big Idea About Genes and Heredity?" prompt
- "Will I Have a Widow's Peak?" simulation
- "Widow's Peak Punnett Squares" activity sheet
- "Freckles or No Freckles?" activity sheet
- "The Genetic Game of Life" lab sheet
- "The Genetic Game of Life Game Cards"
- "Blogging About Sickle Cell Anemia" prompt and checklist
- "Inherited Genetic Disorder Research Guide"
- "Mock Medical Conference Poster Guide"

## SAFETY CONSIDERATIONS

- Have students wear goggles, gloves, and aprons during all phases of the investigation, including during setup and cleanup.
- Review the safety data sheet for isopropyl alcohol (see Appendix 4) and safety information for dish soap.
- Remind students to use caution when working with glassware or plasticware because broken pieces can puncture skin.
- Remind students not to eat the strawberries or any of the other materials used in the investigation.
- Wash hands with soap and water after completing the investigation.

## LEARNING-CYCLE INQUIRY

### Engage

In this phase, students read and discuss an article about how genetic testing provides new options for treating diseases. In discussion, students will identify key words and concepts that will be investigated in subsequent phases of the unit.

1. Distribute copies of the article, "Doctors Get a Jump on Disease With Genetic Testing." (*Note:* This article and several adapted versions [higher and lower reading levels] are available on the NewsELA website [*www.newsela.com*]. Teachers can register for a free account, which allows you to save articles and assign them to classes of students. You may opt to do so and have your students read online, instead of printing copies of the article.)
2. Ask students to read the article through once to themselves, circling or noting unknown words. You can support students of all reading levels by distributing adapted versions of the article or allowing students to read in pairs or in a teacher-directed small group.
3. When students have finished reading the article, ask them to discuss the following questions in pairs:
    - What do you think this article was mainly about?
    - What part of the article did you find most interesting? Why?
    - What are some of the benefits of genetic testing?
    - What are some of the negative consequences of genetic testing?
    - From what you have read, what is your opinion of genetic testing?
4. Prepare for the next step of the activity by organizing five stations for students to share ideas in writing. These stations can take many forms: five pieces of butcher paper with one of the questions written at the top of each paper, five dry-erase

boards with the question written at the top, or one space for each of the five questions where the sticky notes could be posted.

5. Ask the students to select three of the questions from the list to respond to in writing. Allow students approximately five minutes to compose brief answers by moving from station to station to write or by writing on sticky notes and posting them underneath each question.
6. Conduct a class discussion in which you ask a few students to share their responses to each question.
7. Next, ask the students to share the unfamiliar words they circled. Create a list that can be posted in a prominent location for the duration of the unit. Although student responses will vary, we anticipate (and recommend) making sure the following terms make the list: *genes, DNA, extracted, mutant gene, human genome,* and *inherited.*
8. Explain to students that to really understand how genetic testing is used to treat diseases, they will need to learn more about how traits (characteristics) are passed from parents to children. Introduce the unit question, *How can diseases be inherited?*

***Assess this phase:*** Students' participation in individual reading and pair discussion serves as one source of formative assessment for this phase. A second source of formative assessment comes from students' written responses to the questions posed and their contributions to the class discussion. At this time, accept all student ideas, while noting possible misconceptions and areas of confusion.

## Explore I

### PART I

In this phase, students will build background knowledge about DNA by extracting DNA from strawberries and from nonfiction text and video. They will also read nonfiction text to connect DNA to the physical traits they observe in themselves and others.

QR CODE 6

***Advance preparation:*** Prepare for this phase by assembling materials for the DNA extraction, printing the directions for the DNA extraction *(http://1.usa.gov/1RCfv8P;* QR code 6*)* and copying the "Defining DNA" document.

1. Direct student attention to the abbreviation *DNA* on the list. Ask students to return to the article to find the author's explanation of DNA. They should find this sentence: "DNA contains the instructions for how the human body works and is passed on from parents to children." Explain that this sentence will serve as a temporary definition until students have learned more. Post the definition

in a prominent location. Also post the definition of inherited (passed on from parents to children).

2. Ask students to return to the second paragraph of the article to find how doctors were able to examine Denis Keegan's DNA. They should answer that doctors extracted the DNA from his blood.
3. Ask students if they know the meaning of the word *extract*. If not, share that it means to remove or take out. Add this definition to the vocabulary list. Explain to students that they will extract DNA themselves, not from their blood, as in the case of Denis Keegan, but from strawberries.
4. Play the video "How to Extract DNA From Strawberries" *(http://bit.ly/1pO4C92;* QR code 7*)*, and discuss the procedure with students. Next, distribute the directions and review the procedure. Be sure to review safety guidelines and clarify any points that are confusing before allowing students to begin.

QR CODE 7

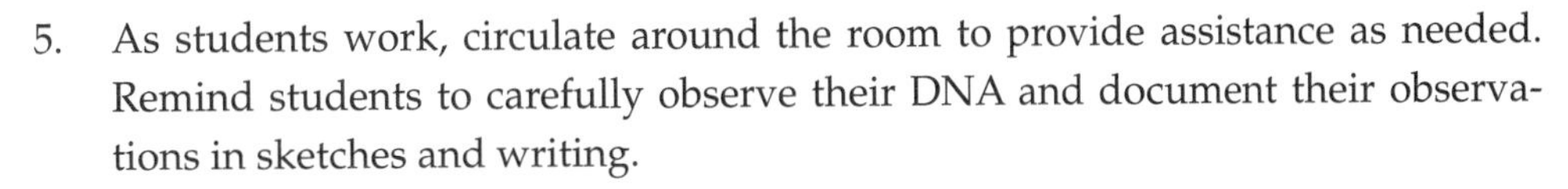

5. As students work, circulate around the room to provide assistance as needed. Remind students to carefully observe their DNA and document their observations in sketches and writing.
6. Next, have students read the article "What Is DNA?," and ask them to jot down notes on the "Defining DNA" graphic organizer as they read. Students will use their observations, as well as information from the text, to develop their own definition of DNA.
7. Discuss student definitions and come to a consensus on a collective definition. Post this on the vocabulary list created earlier in the unit.
8. Ask students, "What does DNA have to do with the traits we observe in humans?" Allow students to think-pair-share, and then distribute copies of the article "Connecting DNA to Heredity." Circulate around the room as students read, providing support as needed.
9. After students have read the article, ask them to identify the significant ideas (big ideas) of the article and record them on "What's the Big Idea About DNA and Heredity?" Ask students, "What does this article tell us about DNA and heredity? What big ideas does the author want us to remember?" Through discussion, compile a list that has the following information:

   - All cells have DNA.
   - DNA forms chromosomes.
   - Genes are sections of chromosomes.
   - Genes contain instructions for making proteins.

- One or more genes control traits.
- Different forms of genes are called *alleles.*
- The combination of alleles determines what trait you have.
- You receive one copy of a gene from your mother and one from your father.

***Assess this phase:*** Student work on the graphic organizers and contributions during the class discussion serve as formative assessment for this phase. At this point in the unit, students should demonstrate a developing understanding of the concepts introduced in the articles. They may still have questions about what they have read and may need assistance with vocabulary terms. These concepts will continue to be developed in the next phase. If students are struggling to comprehend these articles, spend more time here before moving on.

### PART II

In the second part of this phase, students will collect data on the genotypes and phenotypes associated with the widow's peak trait. This data should approximate a monohybrid cross. Students will connect the concepts of genotype and phenotype through discussion.

***Advance preparation:*** Prepare for this phase by labeling two paper bags "Parent 1" and "Parent 2." Cut small squares of paper and write "H" or "h" on each. Each parent has a heterozygous (Hh) genotype (widow's peak phenotype), so the bags should contain an equal number of each allele. We recommend having enough slips of paper in each bag so that students feel they are choosing at random—perhaps 10 H and 10 h alleles in each bag. Also prepare a tally chart with three columns (labeled "two dominant alleles," "two recessive alleles," and "one dominant/one recessive allele") on the board or on butcher paper. (*Note:* Students may question why you are using the letter *h* to represent the alleles of this trait. Explain to students that although *w* or *p* would make more sense in terms of representing the trait, those two letters look very similar in their uppercase and lowercase forms and would be hard to distinguish on the slips of paper. *H* and *h* are more easily distinguished from one another and can represent the word *hairline.*)

1. Introduce the widow's peak trait to students. Show them pictures to illustrate the presence and absence of a widow's peak. Ask students to partner up and determine whether they have a widow's peak.

2. Explain to students that a widow's peak, like some other traits, is controlled by one gene. Explain that there are two versions, or *alleles,* of this gene. Introduce the two alleles, H and h, and label them as dominant (H) and recessive (h). Explain that the combination of alleles is called the *genotype.* Introduce and record the three possible genotypes, labeling them as two dominant alleles (HH), two recessive alleles (hh), and one dominant and one recessive allele (Hh). Keep these genotypes posted in a prominent place for the duration of the simulation.

3. Remind students that an individual inherits one allele from each parent. Explain that they will simulate this process by choosing one allele from two paper bags. They will repeat this for several rounds, recording their data and discussing as they go. Distribute copies of the "Will I Have a Widow's Peak?" data sheet, or direct students to create similar data tables in their science notebooks.

4. To begin round 1, call the first student up to the front of the classroom. Shake each bag and have a student select one slip of paper from each bag without looking. The student should share which alleles he or she selected; place the slip of paper back in the bag; state if he or she selected two dominant alleles, two recessive alleles, or one dominant and one recessive allele; and then return to his or her seat to record the data. Add a tally mark to the chart in the appropriate column. Repeat this step for each student, making a show of shaking the bags up each time. To give students sufficient practice, we recommend allowing each student to repeat the process a second time.

5. Before you begin round 2, introduce the terms *homozygous* and *heterozygous* to students. Explain that homozygous means that an individual has two copies of the same allele, and heterozygous means that an individual has two different alleles. Add the terms *homozygous* and *heterozygous* to the column labels on the tally chart.

6. Begin round 2 by having students come up one at a time again to select alleles from the paper bags. This time, have them share the alleles and if their genotype is homozygous or heterozygous. Again, students should record data on their worksheet while you tally the genotypes on the class chart.

7. Before you begin round 3, introduce the terms *homozygous dominant* and *homozygous recessive* to students. Explain that homozygous dominant refers to an individual with two dominant alleles, whereas homozygous recessive refers to an individual with two recessive alleles. Review that an individual with one dominant and one recessive allele is referred to as *heterozygous*. Modify the labels on the tally chart to reflect this new vocabulary.

8. Begin round 3 by having students come up one at a time again to select alleles from the paper bags. This time, have them share the alleles and if their genotype is homozygous dominant, homozygous recessive, or heterozygous. Again, students should record data on their worksheet while you tally the genotypes on the class chart.

9. Before you begin round 4, introduce the term *phenotype*. Explain to students that the phenotype is the physical expression of the gene, so in this case, the phenotype is either having a widow's peak or not having a widow's peak. Discuss that because having a window's peak is a dominant trait, an individual needs only

one copy of the dominant allele to have that phenotype. Add "widow's peak" or "no widow's peak" to each column on the tally chart.

10. Begin round 4 by having students come up one at a time again to select alleles from the paper bags. This time, have them share the alleles, the genotype (homozygous dominant, homozygous recessive, or heterozygous), and the phenotype (widow's peak or no widow's peak). Again, students should record data on their worksheet while you tally the genotypes on the class chart.
11. Draw students' attention to the tally chart on the board. Total each column and express each genotype as a fraction and a percent. (*Note:* We anticipate that with a sufficient data set, the percentages of each genotype should approach that of a traditional monohybrid cross: 25% homozygous dominant, 50% heterozygous, and 25% homozygous recessive.)
12. Ask students to tell which phenotype each genotype refers to. Label homozygous dominant and heterozygous as "widow's peak," and label homozygous recessive as "no widow's peak." Add the homozygous dominant and heterozygous tallies together, and express the total as a fraction and a percent. (*Note:* The percentages should approach the expected 75% widow's peak: 25% no widow's peak ratio.)
13. Share with students that both parents have a widow's peak. Pose the following question to students: "If both parents have a widow's peak, how are they able to have children with no widow's peak?" Students should answer this question in a quick write. (*Note:* This response is considered a type of formative assessment. Reassure students that their best ideas at this time are sufficient.)
14. Once students have finished their responses, have them read the article "Solving a Genetic Mystery." Before reading, set the purpose that they are reading to find the answer to the question, "If both parents have a widow's peak, how are they able to have children with no widow's peak?" You may read the article aloud with the class, or have students read independently or in groups. Choose the setting that will best support your students with this text.
15. Discuss the article "Solving a Genetic Mystery" with students. Review vocabulary and main ideas from the article and model the process of completing a Punnett square for Mendel's pea plants. Ask students to use vocabulary such as *homozygous dominant, homozygous recessive,* and *heterozygous* as you calculate the expected frequencies of the various genotypes and phenotypes. During this discussion, explain to students that Punnett squares provide expected frequencies and that, in real life, the frequencies may not be exactly what is predicted.
16. Have students complete the "Widow's Peak Punnett Square" worksheet. Walk around and provide assistance as students work.

17. When students have completed the worksheet, discuss their findings and their written responses to the final question. Focus student attention on the monohybrid cross (problem 4) and review that the frequencies are only predictions. It will be helpful to remind students of the previous activity, in which the paper bags were shaken up in between each student's selection. Review the frequencies obtained in the activity, and discuss how they approached, but were not exactly, 75% widow's peak and 25% no widow's peak. It is important that you stress that Punnett squares help *estimate* (not predict) the chance a trait may be realized in offspring.
18. Explain to students that a type of diagram called a *pedigree* is one way to show how traits are passed from one generation to the next. Lead students through the process of constructing a pedigree for the parents in problem 4 on the "Widow's Peak Punnett Squares" worksheet, imagining that the parents had two children. Students should construct this pedigree in the space on their worksheet as you demonstrate on the board.

*Assess this phase:* Student participation in the simulation, completed data sheets, written responses to the reflection, and Punnett squares all provide important sources of formative assessment. Students should be able to correctly label genotypes as homozygous dominant, homozygous recessive, or heterozygous. They should also be able to match the correct phenotype to each genotype and use Punnett squares to calculate expected frequencies for offspring. If students are having difficulty with these concepts, return to the simulation and repeat the activities in a more teacher-directed manner.

## Explain I

In this phase, students will create Punnett squares, write explanations, and construct a simple pedigree to demonstrate their understanding of dominant and recessive traits.

1. Give students the handout "Freckles or No Freckles?" and review the directions.
2. Explain to students that they should use what they have learned so far to answer the questions. (*Note:* Students should have access to all unit materials while they complete this assignment.)

*Assess this phase:* Student work on the "Freckles or No Freckles?" handout is one piece of summative assessment for this unit. If students do not demonstrate mastery of concepts on this assessment, return to the Explore I phase and reteach as necessary before proceeding.

## Explore II

In this phase, students will investigate sickle cell anemia to learn how a disease can be inherited.

1. Return to the article "Doctors Get a Jump on Disease With Genetic Testing." Read the first three paragraphs aloud with students, drawing their attention to the term *mutant gene.* Ask students to think-pair-share about what this term might mean.
2. Read aloud the article "Mutations and Disease" to students. After reading the article, define *mutation* (a change in the genetic code of a gene), and post it on the vocabulary board. Ask students to share something they learned about mutations from the article. Record these ideas on the board or on chart paper. Re-read and use guiding questions to help students focus on the following points:
   - Mutations are common; everyone has mutations in his or her DNA.
   - Few mutations are harmful.
   - Some mutations cause disease.
   - Most inherited genetic diseases are recessive.
3. Explain to students that they will study the heredity of one family to learn more about mutations and disease. Show students the video "A Mutation Story." Spend some time after the video discussing sickle cell anemia. Help focus student attention on the fact that the sickle cell mutation is a recessive allele and that two copies are needed for an individual to have sickle cell anemia.
4. Distribute copies of "The Genetic Game of Life" to students. Explain that they will play a game to explore the inheritance of sickle cell anemia through the family and will represent this information in a pedigree. (*Note:* We recommend having the game cards in a large bowl in the front of the classroom. Call one group up at a time to select its cards, ensuring that the group members return their cards to the bowl before another group has a turn.)
5. When students have finished their pedigrees, post them around the classroom. Conduct a gallery walk in which students observe each other's pedigrees. Discuss student observations, reiterating the idea that the inheritance of traits is largely up to chance and that, although we can predict expected frequencies, we can't expect real-life results to necessarily match these frequencies.

***Assess this phase:*** Student participation and completion of the pedigree provide a means of formative assessment. By the end of this phase, students should be comfortable using correct terminology (*genotype, phenotype,* etc.) and should demonstrate an understanding of the mechanisms behind the inheritance of one trait.

## Explain II

In this phase, students will write a blog post to answer a variation of the question, *How can diseases be inherited?*

1. Distribute the "Blogging About Sickle Cell Anemia" handout. Read over the RAFT (role of the writer, audience, format, and topic) writing prompt with students (see page 344 in Appendix 2 for an explanation of the RAFT strategy):

   You are a **science blogger** for a student news organization, What Do You Know? (WDYK). Your editor has asked you to write a **blog post** telling the story of a couple, Fatu and Shakashare, and their **family history of the disease sickle cell anemia.** Your editor has given you some notes from an interview with the couple for use in your blog post. Your readers are mostly **middle and high school students,** so your editor would like you to also include an **explanation of how the disease is inherited.**

2. Students should have access to all unit materials as they complete the graphic organizer. Their primary source of information should be their data table and pedigree from the previous phase. Collect and review the graphic organizers before allowing students to begin writing their blog posts. This is an important, final piece of formative assessment.
3. Remind students to use the "Blogging About Sickle Cell Anemia" checklist as they write. Student blog posts can take many forms, ranging from pencil-and-paper documents to published posts on a class blog. Consider the needs and interests of your students, your technological capabilities, and the time frame as you determine what form to use. Regardless of the form used, provide time for students to read each other's work and leave comments, simulating the interactive nature of a blog.

***Assess this phase:*** The "Blogging About Sickle Cell Anemia" graphic organizer serves as a final formative assessment for this unit. It is essential that students demonstrate an understanding of concepts before proceeding to write their actual blog post. If students do not show mastery of concepts, return to the Explore phase and re-teach before continuing with the unit.

## Expand

In this phase, students will research and share information about heritable diseases.

1. Divide students into small groups, and assign each group an inherited genetic disorder: cystic fibrosis, phenylketonuria, achondroplasia, Tay-Sachs disease, or hemophilia. (*Note:* Hemophilia is a sex-linked trait, so it is an excellent topic for a group of students needing extension.)

2. Explain to students that they will be participating in a mock medical conference on inherited diseases. Each group is responsible for creating a miniature poster (using a manila file folder) to share information about the disease and how it is inherited. Distribute the "Inherited Genetic Disorder Research Guide" handout to students and allow them to begin researching.
3. Once students have collected sufficient information, distribute the "Mock Medical Conference Poster Guide" handout to groups and allow them to begin work on their posters.
4. When groups have completed their posters, conduct the mock conference. You may opt to invite other teachers, parents, or administrators to listen to groups present their information.

*Assess this phase:* Each group's poster serves as a final summative assessment for the unit. Students should be able to demonstrate, in writing and speaking, major concepts and vocabulary related to the inheritance of traits.

## REFERENCES

National Governors Association Center for Best Practices and Council of Chief State School Officers (NGAC and CCSSO). 2010. *Common core state standards.* Washington, DC: NGAC and CCSSO.

NGSS Lead States. 2013. *Next Generation Science Standards: For states, by states.* Washington, DC: National Academies Press. *www.nextgenscience.org/next-generation-science-standards.*

Radford, A., and J. A. Bird-Stewart. 1982. Teaching genetics in schools. *Journal of Biological Education* 16 (3): 177–180.

Shaw, K. R. M., K. Van Horne, H. Zhang, and J. Boughman. 2008. Essay contest reveals misconceptions of high school students in genetic content. *Genetics* 178 (3): 1157–1168.

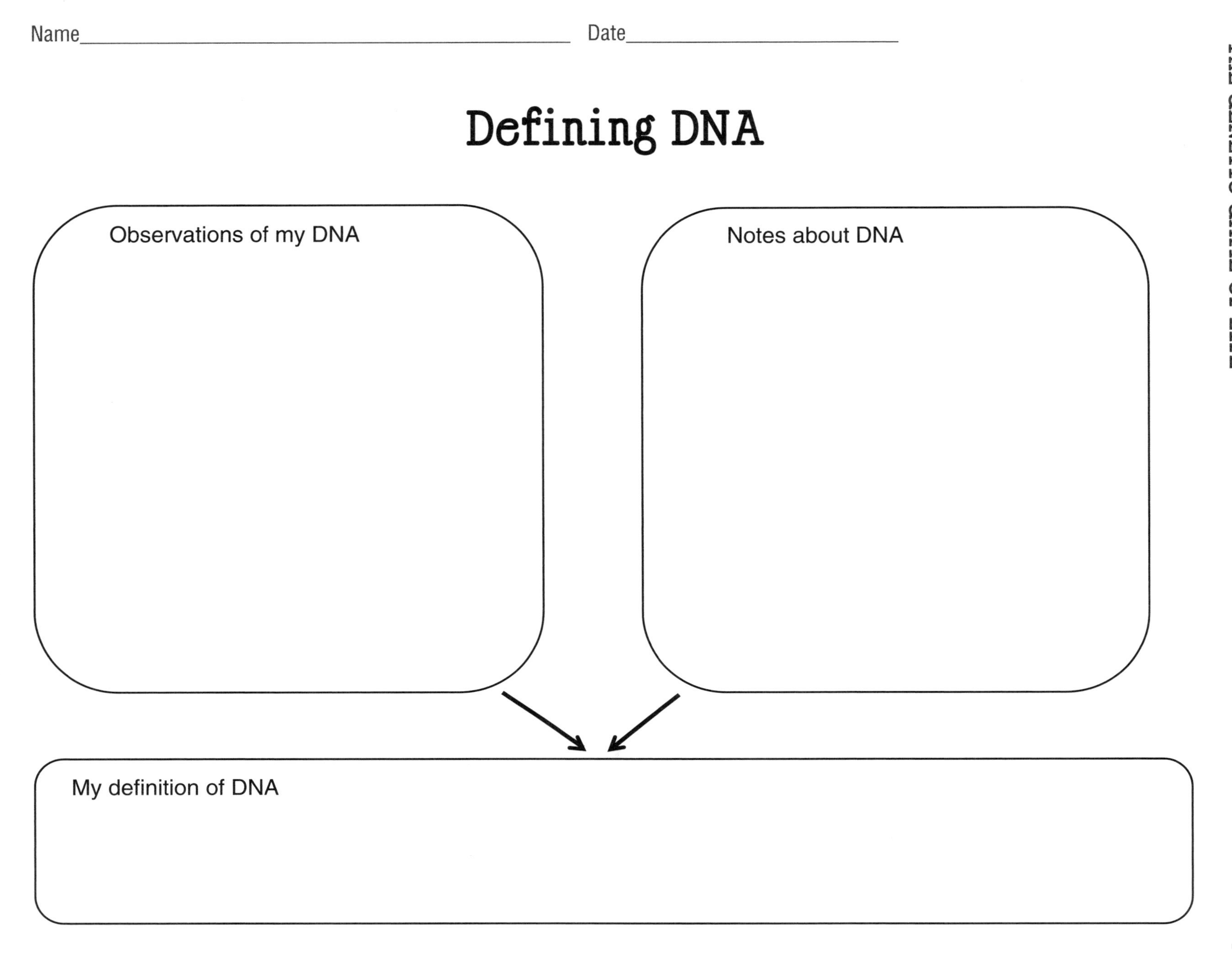
Name
Date
Defining DNA
Observations of my DNA
Notes about DNA
My definition of DNA

Name_______________________________________ Date_______________________

# Connecting DNA to Heredity

You know that all cells contain DNA and that DNA wraps together to form chromosomes. Most cells in your body have 23 pairs, or 46 chromosomes. One-half of these chromosomes were passed on to you from your father. The other half of these chromosomes were passed on to you from your mother. The combination of your chromosomes makes you who you are. But what do these chromosomes have to do with traits such as eye color and hair color? How do these chromosomes make you look like your mother or your father—or a combination of the two?

To answer these questions, we need to look more closely at chromosomes. Each chromosome is made up of smaller sections called **genes.** Each gene is a recipe, or set of directions, for making a certain kind of protein. These proteins are what make you have certain traits. Eye color, hair color, height, and more all come down to the kinds of proteins your cells produce.

Some traits are coded by one gene. Other traits are coded by many genes. Freckles and dimples are two examples of human traits that are coded for by one gene. Eye color, hair color, and height are all examples of traits that are coded by many genes.

Each gene comes in different forms. These forms are called **alleles.** For example, consider the trait of having freckles or no freckles. One allele of the gene codes for freckles; the other allele of the gene codes for no freckles. The combination of the alleles determines whether someone will have freckles.

One copy of each gene is passed down to you by your mother, and one copy is passed down to you by your father. If the trait is produced by one gene, you will have either your mother's trait or your father's trait, depending on the combination of alleles you received. If the trait is produced by more than one gene, the traits you inherit will be more complicated.

Understanding heredity can be difficult, but it's amazing to think that it all comes down to tiny sections of DNA—our genes.

Name________________________ Date______________

# What's the Big Idea About Genes and Heredity?

Re-read "Connecting DNA to Heredity" and look for the big ideas in the text. Ask yourself the following question: What ideas does the author want me to remember? Record these big ideas as a list below.

Name________________________ Date________________

# Will I Have a Widow's Peak?

## ROUND 1

| Allele One | Allele Two | Genotype Description (two dominant alleles, two recessive alleles, one dominant or one recessive allele) |
|---|---|---|
| | | |
| | | |

## ROUND 2

| Allele One | Allele Two | Genotype Description (two dominant alleles, two recessive alleles, one dominant or one recessive allele) | Genotype Description (homozygous or heterozygous) |
|---|---|---|---|
| | | | |
| | | | |

## ROUND 3

| Allele One | Allele Two | Genotype Description (two dominant alleles, two recessive alleles, one dominant or one recessive allele) | Genotype Description (homozygous dominant, homozygous recessive, or heterozygous) |
|---|---|---|---|
| | | | |
| | | | |

## ROUND 4

| Allele One | Allele Two | Genotype Description (two dominant alleles, two recessive alleles, one dominant or one recessive allele) | Genotype Description (homozygous dominant, homozygous recessive, or heterozygous) | Phenotype (widow's peak or no widow's peak) |
|---|---|---|---|---|
| | | | | |
| | | | | |

In our simulation, ________ % of the children had a widow's peak, and ________ % of the children did not have a widow's peak.

Name______________________________ Date____________________

Will I Have a Widow's Peak? (*continued*)

## REFLECTION

If both parents have a widow's peak, how are they able to have children with no widow's peak?

Name______________________ Date______________

# Widow's Peak Punnett Squares

Complete the Punnett square for each situation.

1. Parents' Genotypes: HH and hh
   Parents' Phenotypes: ______________ and ______________

| | |
|---|---|
| | |
| | |

Children's Genotypes:

Children's Phenotypes:

2. Parents' Genotypes: Hh and HH
   Parents' Phenotypes: ______________ and ______________

| | |
|---|---|
| | |
| | |

Children's Genotypes:

Children's Phenotypes:

3. Parents' Genotypes: Hh and hh
   Parents' Phenotypes: ______________ and ______________

| | |
|---|---|
| | |
| | |

Children's Genotypes:

Children's Phenotypes:

Name______________________________ Date________________

## Widow's Peak Punnett Squares (*continued*)

4. Parents' Genotypes: Hh and Hh
   Parents' Phenotypes: ________________ and ______________

| | |
|---|---|
| | |
| | |

Children's Genotypes:

Children's Phenotypes:

Which set of parents gave results that were similar to what you found in your class simulation?

______________________________________________________________

______________________________________________________________

______________________________________________________________

______________________________________________________________

______________________________________________________________

______________________________________________________________

## PEDIGREE DRAWING

Name______________________________ Date__________________

# Freckles or No Freckles?

Having freckles is a trait controlled by a single gene. There are two alleles of the gene: freckles (E) and no freckles (e). Having freckles is the dominant allele, and not having freckles is the recessive allele. Fred has freckles. His genotype is Ee. His wife, Peyton, does not have freckles. Her genotype is ee.

1. Create a Punnett square to show the possible genotypes of their children.

2. What is the chance of their children having freckles? Explain your thinking.

3. Imagine that Fred and Peyton have a daughter with freckles and a son without freckles. Draw a pedigree to show their family's genotypes.

Name________________________________ Date________________

# The Genetic Game of Life

## BACKGROUND INFORMATION

In the video "A Mutation Story," you learned about sickle cell anemia and how it is based on a genetic mutation. You met a couple, Fatu and Shakasare, who both carry one allele for sickle cell anemia (Aa). We say that they have sickle cell trait. Although they do not have the disease, sickle cell trait provides a surprising benefit: It helps protect them from malaria. In this game, you will predict the genotypes and phenotypes of their children and draw a pedigree to show their family.

QR CODE: "A MUTATION STORY"

## PROCEDURE

1. Roll the dice to determine how many children the couple will have.
2. Draw this number of cards to learn the genotype of each child. Record this information in the data table below. Return the cards to the bowl immediately so that another group can take a turn.
3. Determine each child's phenotype from the genotype cards that you selected at random. Record this information next to each genotype below.
4. On a separate piece of paper, create a pedigree that shows the family. Use the key shown below as you create your pedigree.

## DATA

Number of children ____________

| Children's Genotypes | Children's Phenotypes |
|---|---|
| | |
| | |
| | |
| | |
| | |
| | |

Name______________________________ Date__________________

## The Genetic Game of Life (*continued*)

Use the information from the game and the key to create a pedigree for the family. The parents, Fatu and Shakashare, have been added for you.

**Pedigree Key**

| Symbol | Meaning | Symbol | Meaning |
|---|---|---|---|
| ○ | Normal female | □ | Normal male |
| ◐ | Carrier female (sickle cell trait) | ◧ | Carrier male (sickle cell trait) |
| ● | Affected female (sickle cell anemia) | ■ | Affected male (sickle cell anemia) |

Name________________________ Date________________

# The Genetic Game of Life Game Cards

| | |
|---|---|
| Child's Genotype<br>AA | Child's Genotype<br>AA |
| Child's Genotype<br>AA | Child's Genotype<br>AA |
| Child's Genotype<br>AA | Child's Genotype<br>aa |
| Child's Genotype<br>aa | Child's Genotype<br>aa |
| Child's Genotype<br>aa | Child's Genotype<br>aa |

Name________________________________ Date________________

## The Genetic Game of Life Cards *(continued)*

| | |
|---|---|
| Child's Genotype<br><br>Aa | Child's Genotype<br><br>Aa |
| Child's Genotype<br><br>Aa | Child's Genotype<br><br>Aa |
| Child's Genotype<br><br>Aa | Child's Genotype<br><br>Aa |
| Child's Genotype<br><br>Aa | Child's Genotype<br><br>Aa |
| Child's Genotype<br><br>Aa | Child's Genotype<br><br>Aa |

Name________________________________________ Date______________________

# Blogging About Sickle Cell Anemia

You are a **science blogger** for a student news organization, What Do You Know? (WDYK). Your editor has asked you to write a **blog post** telling the story of a couple, Fatu and Shakashare, and their **family history of the disease sickle cell anemia.** Your editor has given you some notes from an interview with the couple for use in your blog post. Your readers are mostly **middle and high school students**, so your editor would like you to also include an **explanation of how the disease is inherited.**

Use the space below to plan your blog post.

| Information about the family's history of sickle cell anemia | Information about how sickle cell anemia is inherited |
| --- | --- |
| | |

Name________________________________ Date__________________

## Blogging About Sickle Cell Anemia (*continued*)

Your editor gave you the checklist below. Use it to make sure you include all the necessary information in your post.

1. Family history of sickle cell anemia
   - ☐ Mother (Fatu)
   - ☐ Father (Shakashare)
   - ☐ Children

2. How sickle cell anemia is inherited
   - ☐ Dominant and recessive alleles
   - ☐ Genotypes and phenotypes
   - ☐ Expected frequencies of each genotype and phenotype in this family

3. Other information
   - ☐ Vocabulary: *homozygous, heterozygous,* and *gene*

Name______________________________ Date____________________

# Inherited Genetic Disorder Research Guide

Disease ______________________________________

| About the disease (symptoms, characteristics) | How the disease is inherited |
| --- | --- |
| | |
| How the disease is treated | |
| Sources | |

Name______________________________ Date__________________

# Mock Medical Conference Poster Guide

You will create your poster using a manila folder. Your poster should include the following information:

- Name of the disease
- Description of the disease's characteristics and symptoms
- Information about how the disease is inherited
- A pedigree showing a family that has at least one member (parent or child) affected by the disease
- Information about how the disease is treated
- A bibliography (can be on a separate piece of paper or on the back of the poster)

## Chapter 8

# Seriously ... That's Where the Mass of a Tree Comes From?

## OVERVIEW

In this unit, students conduct an investigation to answer the question, *Where does the mass of a plant come from?* Ultimately, students determine carbon dioxide accounts for the dry mass of a plant. Along the way, they discover that photosynthesis is the process through which plants make food (glucose).

This unit assumes that students know and can demonstrate the basic needs and structures of plants, cellular structures of plants, and how to use lab equipment such as balances and graduated cylinders.

## OBJECTIVES

- Identify photosynthesis as the process through which plants make glucose.
- Determine that most of the mass of a plant comes from carbon dioxide.
- Explain that plants use glucose for energy and raw materials for plant growth and development.
- Analyze data and draw logical conclusions.
- Synthesize information gathered through investigations and text.

## STANDARDS ALIGNMENT

### *Next Generation Science Standards* (NGSS Lead States 2013)

#### LS1.C: ORGANIZATION FOR MATTER AND ENERGY FLOW IN ORGANISMS

- Plants, algae (including phytoplankton), and many microorganisms use the energy from light to make sugars (food) from carbon dioxide from the atmosphere and water through the process of photosynthesis, which also releases oxygen. These sugars can be used immediately or stored for growth or later use. (MS-LS1-6)

#### PS3.D: ENERGY IN CHEMICAL PROCESSES AND EVERYDAY LIFE

- The chemical reaction by which plants produce complex food molecules (sugars) requires an energy input (i.e., from sunlight) to occur. In this reaction, carbon dioxide and water combine to form carbon-based organic molecules and release oxygen. (secondary to MS-LS1-6)

### Common Core State Standards, English Language Arts (NGAC and CCSSO 2010)

#### GRADES 6–12 LITERACY IN HISTORY/SOCIAL STUDIES, SCIENCE, AND TECHNICAL SUBJECTS

- CCSS.ELA-LITERACY.RST.6-8.1: Cite specific textual evidence to support analysis of science and technical texts.
- CCSS.ELA-LITERACY.RST.6-8.3: Follow precisely a multistep procedure when carrying out experiments, taking measurements, or performing technical tasks.
- CCSS.ELA-LITERACY.RST.6-8.9: Compare and contrast the information gained from experiments, simulations, video, or multimedia sources with that gained from reading a text on the same topic.

## TIME FRAME

- Four weeks for plants to grow (one day to plant the seeds, followed by short periods of time for watering)
- Five to six 45-minute class periods

## SCIENTIFIC BACKGROUND INFORMATION

Photosynthesis is a critical biological process that supports heterotrophic life on Earth. Photosynthesis produces glucose that forms the base of the food chain and oxygen for aerobic organisms. It is safe to say that without photosynthesis, most life on Earth would not exist.

Photosynthesis is a complex biochemical process that occurs in plants, algae, some protists, and certain prokaryotes. At the simplest level, photosynthesis is the process of chemically combining carbon dioxide ($CO_2$) and water ($H_2O$) to produce glucose ($C_6H_{12}O_6$) and oxygen ($O_2$). The process is powered by energy from the Sun. The chemical formula for photosynthesis represents the reaction that converts compounds from inorganic to organic (see Figure 8.1).

#### FIGURE 8.1. THE CHEMICAL FORMULA FOR PHOTOSYNTHESIS

$$6\,CO_2 + 6\,H_2O \longrightarrow C_6H_{12}O_6 + 6O_2$$

Photosynthesis occurs in leaves at the cellular level. Plant cells have specialized organelles, called *chloroplasts,* which function in photosynthesis. Chloroplasts absorb light energy, split water, release oxygen, produce cellular energy (ATP, or adenosine triphosphate) and NADPH (an electron carrier), and produce glucose. The light energy chloroplasts split is captured by chlorophyll, a light-absorbing molecule found in the chloroplast. Chlorophyll absorbs all wavelengths of light *except* green. Green light is reflected, which is why most leaves appear green to us.

The work of chloroplasts takes place in two phases commonly referred to as the *light reaction* and the *dark reaction*. In the light reaction, sunlight is absorbed, water is split into hydrogen and oxygen, oxygen is released, and—through a series of steps—ATP and NADPH are produced. This process is also called the *light dependent reaction*. In the dark reaction, carbon dioxide is captured and chemically combined with the hydrogen from water to produce glucose. This stage is driven by the ATP and NADPH. The ATP is consumed and NADP is recycled (the hydrogen is released and used in the glucose). Light is not required for the dark reaction; thus, it is also called the *light-independent reaction*. The light-independent reaction can occur in the presence of light, but light is not required for this part of photosynthesis.

This description of the work of chloroplasts is greatly simplified. Photosynthesis is a complex process involving many steps. The process was intentionally oversimplified in our discussion to focus on the big idea that middle school students should understand. It is important for them to understand that chloroplasts in plant cells do the work of photosynthesis and that chlorophyll is the molecule that captures sunlight. That's really all they need to know about the biochemical process of photosynthesis. What they really need to know and deeply understand is that this process is critical to life on Earth. They should know that the process is powered by energy from the Sun, understand what the raw materials are and how they enter the plant, and learn that plants use the glucose they make for energy and structural components of the plant (i.e., starches and cellulose).

The raw materials (sunlight, water, and carbon dioxide) are readily available in most places. Where these resources are in short supply (water in the desert for instance), plants have adapted. Generally speaking, water is absorbed through the roots of plants and carried through the plant to the leaves in vascular structures known as *xylem*. Evapotranspiration, the evaporation of water through plant leaves, plays an important role in this process. Sunlight is absorbed by chlorophyll in the chloroplasts of leaves. The carbon dioxide enters the leaf through a cellular structure known as *stomata*. Then, within the chloroplast, photosynthesis takes place.

An interesting way to consider photosynthesis is from the perspective of where the mass of a plant comes from. It turns out that nearly all of the mass of a plant comes from the carbon dioxide the plant takes in from the atmosphere. That seems surprising and perhaps a bit suspect. Even before doing experiments to demonstrate this, you can take a look at the products of photosynthesis—oxygen and glucose. The oxygen is released, so it does not become part of the plant. The glucose is retained by the plant for its own biological processes. Glucose is $C_6H_{12}O_6$; it has an atomic mass of 180.16 g. Hydrogen composes a mere 6.66% of the atomic mass. Most of the mass of glucose is in the carbon and oxygen that were taken in as carbon dioxide.

Given that plants are the base of food webs, many middle school students think that plants make food for animals. This suggests that plants behave in a sort of philanthropic way. Plants use the glucose they make for energy and structural components of the plant. Animals prey on plants for food. This type of predation is known as *herbivory*. Plants do form the base of food webs, no question about it, but the process of photosynthesis supports the life processes of plants.

## MISCONCEPTIONS

Misconceptions about photosynthesis indicate that students often think that plants make food and release oxygen for the benefit of people and other animals. Students often view photosynthesis as a kind of inverse respiration, and they rarely understand that the mass of a plant, even a full-grown tree, comes from the carbon dioxide taken in through photosynthesis. In *A Private Universe*, graduates at Harvard University's commencement were not able to identify carbon dioxide as the source of a plant's mass. Other common misconceptions are presented in Table 8.1.

### TABLE 8.1. COMMON MISCONCEPTIONS ABOUT PHOTOSYNTHESIS

| Common Misconception | Scientifically Accurate Concept |
|---|---|
| Plants get their food from the soil. | Plants produce glucose through photosynthesis. The glucose is a source of energy and structural components of the plant. Plants take in minerals from soil. |
| Photosynthesis is something plants do for people and animals (applies to production of food and oxygen). | Photosynthesis does produce glucose and release oxygen. Plants use the glucose. They also use oxygen for cellular respiration. People and other animals prey on plants as a food source. |
| Weight increase in plants is from soil, water, or both. | Most of the dry weight of a plant comes from carbon dioxide. |
| Chlorophyll attracts light. | Chlorophyll is a light-absorbing molecule. Sunlight is abundant; there is no need for a molecule that "attracts" light. Plants use only a small portion of available sunlight. |
| • Plants get their energy from the Sun.<br>• Heat from the Sun is used as energy for photosynthesis. | Plants absorb sunlight, which drives the formation of the molecules that will be used to power the production of glucose. Heat from the Sun warms plants, but it is not a source of energy for the plant. |
| Plants breathe in carbon dioxide and breathe out oxygen. | Through photosynthesis, plants absorb carbon dioxide and release oxygen. In cellular respiration, plants take in oxygen and release carbon dioxide, just as other organisms do. |

*Source:* Canal 1999; Driver 1994; Ozay and Oztas 2003.

## NONFICTION TEXTS

"Acorn to Oak Filmed Over an 8 Month Period Time-Lapse" by Neil Bromhall, 2011, *http://bit.ly/20BNPWB.* (QR code 1)

QR CODE 1

In three minutes, this time-lapse video shows the germination of an acorn and growth of an oak sapling over a period of 8 months.

*Ancient Ones: The World of the Old-Growth Douglas Fir* by Barbara Bash (Oakland, CA: Sierra Club Books for Children, 1994).

In this book, Bash describes the old growth forests of the Pacific Northwest. Most of the book focuses on the Douglas fir and the organisms that make it their home.

"Comparison of Features of Oaks," habitat images from *The Nut Trees of Land Between the Lakes,* by Edward W. Chester, Richard J. Jensen, Louis J. Schibig, and Suzanne Simoni (Clarksville, TN: Austin Peay State University, 1987); *http://bit.ly/1K5Wj5U.* (QR code 2)

QR CODE 2

This key to oaks provides the tree's genus and species name and habitat, and also contains images of the tree's leaves, fruit, twigs, buds, and bark.

*Desert Giant: The World of the Saguaro Cactus* by Barbara Bash (Oakland, CA: Sierra Club Books for Children, 1989); published Lexile level 980L.

This book offers beautiful illustrations of the saguaro cactus and the animals that visit it. The book describes the saguaro in great detail and explains the importance of the cactus to the desert ecosystem.

QR CODE 3

"How Plants Use the Glucose" by Melbourne High School, Victoria, Australia, n.d., *http://bit.ly/1PZC76D.* (QR code 3)

This infographic describes how plants use the sugar they make for energy to carry out life processes.

*In the Heart of the Village: The World of the Indian Banyan Tree* by Barbara Bash (Oakland, CA: Sierra Club Books for Children, 1996).

The significance of a large banyan tree in the middle of a small rural village is described in lyrical prose in this beautifully illustrated book.

*Photosynthesis* by Bonnie Juettner (Kidhaven Science Library, 2005); Flesch-Kincaid reading level 6.7, published Lexile level 1010L.

Drawings, diagrams, and photographs help middle-grade readers grasp the importance of photosynthesis and help them gain insight into the process.

*Photosynthesis* by Christine Zuchora-Walske (Edina, MN: ABDO Publishing, 2014); Flesch-Kincaid reading level 5.9, published Lexile level 850L.

This book describes photosynthesis from the cellular to the ecosystem level. The author stresses the importance of photosynthesis to all life on Earth.

"Photosynthetic Cells" Scitable by Nature Education, 2014, *http://bit.ly/1AH0bFK.* (QR code 4)

QR CODE 4

This article provides detailed information about the process of photosynthesis. The information includes pictures of plants, diagrams of cellular structures, and flowcharts illustrating the steps in photosynthesis.

"Plants Are Made From Thin Air?" by Edgerton Center, Massachusetts Institute of Technology, Cambridge, MA, n.d., *http://bit.ly/1SQUKfG. (QR code 5)*

QR CODE 5

This resource describes the ways plants use the glucose they make.

Real Trees 4 Kids! website, The National Christmas Tree Association, last updated 2008, *http://bit.ly/1ix4Rl0.* (QR code 6)

QR CODE 6

This website provides a basic description of photosynthesis that is easy to understand.

*Tree of Life: The World of the African Baobab* by Barbara Bash (Oakland, CA: Sierra Club Books for Children, 1989); published Lexile level 1040L.

In this beautifully illustrated book, Barbara Bash describes the baobab tree in detail and shares some of the African folklore associated with this impressive tree.

*Understanding Photosynthesis With Max Axiom, Super Scientist* by Liam O'Donnell (North Mankato, MN: Capstone Press, 2007); Flesch-Kincaid reading level 5.2, published Lexile level 780L.

In this graphic novel, the story of photosynthesis is told by super scientist Max Axiom. The complexities of photosynthesis are broken down through the combination of colorful panels and the narrative provided by the main character.

QR CODE 7

"What Happens to the Glucose Plants Make During Photosynthesis?" by Indiana University, Bloomington, n.d., *http://bit.ly/1F5XlYG.* (QR code 7)

This interactive animation explains what happens during the light and dark reactions of photosynthesis.

## MATERIALS

- Bean seeds (5 per group)
- Soil
- Cups to grow the seeds in
- Graduated cylinder (1 per group)
- Balances
- Containers for drying soil samples and plants (e.g., cupcake papers)
- Bromothymol blue (BTB; see safety data sheet information in Appendix 4)
- Oven (for use in drying soil and plants)
- Filter papers (preferably coffee filters cut to the correct size)
- Funnels
- 1 jumbo resealable bag
- Small containers for BTB (condiment cups work well)
- Beakers (400 ml) or cups (12 oz.) (1 per group)
- Straws with a small cut near one end
- Test tube and stopper or vial with cap (1 per group)
- Test tubes
- Test tube racks
- Sprigs of *Elodea* or *Cabomba* (available at pet stores and science supply stores)
- Colored cellophane or acetate
- 1% sodium bicarbonate solution (see safety data sheet information in Appendix 4)
- Grow light
- Stop watch
- Tree field guides (e.g., Audubon or Peterson field guides if choosing the field guide option for the Engage phase)
- Goggles (sanitized, indirectly vented chemical splash)
- Nonlatex gloves
- Aprons
- Small watercolor (or similar) paint brushes

## SUPPORTING DOCUMENTS

- "Comparing Masses" lab sheet
- "Plant Mass Investigation and Data Sheet"
- "Photosynthesis: Powering the Planet" prompt
- "Mass From the Air!" lab sheet
- "Speaking of Photosynthesis …" prompt

## SAFETY CONSIDERATIONS

- Have students wear goggles, gloves, and aprons during all phases of the investigation, including during setup and cleanup.
- Review safety information in the safety data sheets for hazardous chemicals (BTB, sodium bicarbonate solution, and some food colorings).
- Remind students to use caution when working with glassware or plasticware, grow light bulbs, and so on. These items could puncture skin if broken.
- Remind students not to eat beans used in investigations because they may have pesticides or herbicides on them.
- Use only a GFI-protected wall electrical outlet when working near water, and plug grow light fixtures into GFI circuits only.
- Use caution when working around drying ovens because they can be hot and burn skin.
- Use soil that is pesticide and herbicide free.
- Wash hands with soap and water after completing the investigation.

## LEARNING-CYCLE INQUIRY

### Engage

The Engage phase of this unit is designed to pique students' curiosity about how a massive tree develops from a relatively small seed. There are two parts to the Engage activities. In the first part, the students begin thinking about the difference between the mass of a tree seed compared to the mass of a mature tree. In the second part, students determine the average mass of a tree seed and then calculate the mass of a mature tree. Several options are presented for the first step. Select the option that will work best for your students.

#### PART I: COMPARING TREE SEEDS TO MATURE TREES

Several options for the Engage phase are presented here. Choose the option that makes the most sense for your students. Consider differentiating instruction by offering several of the activities to the class and allowing them to choose the activity that most appeals to them. Some students could work with field guides, others with the Bash books, and still others with the video. After all students have finished their chosen activity, bring the class together for sharing and discussion.

### Option 1: Time-Lapse Video

1. Show the "Acorn to Oak Filmed Over an 8 Month Period Time-Lapse" video to the entire class.
2. After the video, ask students what they notice about the acorn.
3. Share some images of mature oak trees. There are hundreds of species of oak trees, most of which live in the Northern Hemisphere. Consider selecting several species of oak that grow in your location. The nonfiction text "Comparison of Features of Oaks" is a good place to start. This site includes images of mature oaks, acorns, leaves, bark, and so on.
4. Lead a short classroom discussion about the size of the acorn compared to the size of the mature tree.

### Option 2: Field Guides

1. Divide students into small work groups of three to four. Provide each group with a field guide or set of field guides for trees (e.g., Audubon or Peterson field guides). Instruct students to pick three to four trees from the field guide and record interesting information about each tree. Draw a table like the one shown in Table 8.2 to help students think about how to organize their information. The information must include the average size of the mature tree and the tree's seed. Allow students to determine what additional information they would like to record.

#### TABLE 8.2. SAMPLE DATA TABLE FOR FIELD GUIDE INVESTIGATION

| Tree | Seed Information | Mature Tree Height | Additional Fact 1 | Additional Fact 2 |
|---|---|---|---|---|
| Sugar maple | | | | |
| Red oak | | | | |

2. Ask each group to share what they learned after they are finished gathering their data.
3. Lead a short classroom discussion about the size of the seed compared to the size of the mature tree.

### Option 3: Trees in a Social and Ecological Context

1. Provide small groups of students with copies of Barbara Bash's books about trees: *Ancient Ones: The World of the Old-Growth Douglas Fir; Desert Giant: The World of the Saguaro Cactus; In the Heart of the Village: The World of the Indian Banyan Tree;* and *Tree of Life: The World of the African Baobab.*
2. Ask groups to make note of striking features of trees and their role in communities, folklore, or the ecosystem as they read. In each book, students will find information about the size of the tree and a description of the seed.
3. Hold a class discussion about interesting characteristics of the trees. Be sure to ask for descriptions of the trees' sizes and seeds.

## PART II: DETERMINING THE RATIO OF SEED TO TREE MASS

1. Students should determine the average mass of a tree's seeds. Of the following options, choose the one that works best for your students:
   a. If you have access to acorns or other tree seeds, provide groups of students with 10–15 seeds. Ask students to mass the seeds, and determine the average mass of the seeds. This is a good time to review how to calculate an average. Students may find the "Comparing Masses" worksheet on pages 150–151 helpful.
   b. If seeds are not available, you can gather information about the number of seeds per pound for a wide variety of tree species by visiting seed company sites (e.g., Sheffield's Seed Company [*http://bit.ly/1NRlww1*] and F. W. Schumacher Co. [*http://bit.ly/20s3yN5*]. For this option, students can calculate the average mass of a seed without actually measuring the seeds. (See QR codes 8 and 9.)

QR CODE 8

QR CODE 9

2. Determine the mass of the tree the seeds came from. Of the following options, choose the one that works best for your students:
   a. If there are trees in the area that are the same species as the seeds that were massed, have students measure the circumference of the trees 4.5 feet from the ground. Students will then use the Mass in Kilograms of the Trunk of Hardwood Trees data table on the "Comparing Mass" sheets to determine the approximate mass of the tree they measured.
   b. If trees are not available, refer to online resources such as "What Tree Is It?" *(http://bit.ly/1mmlLWf)* or "eNature" *(http://bit.ly/23K7oAx)*. Students can search for trees by common names on either site. They may need to convert from metric to standard units. If doing so will be difficult for students, consider using an online conversion tool such as "Metric Conversions" *(http://bit.ly/1lNs2Yi)*. (See QR codes 10–12.)

QR CODE 10

QR CODE 11

QR CODE 12

3. Ask students to calculate the ratio of the mass of the tree to the mass of the seed by dividing the mass of the tree by the the average mass of the seeds. This step provides a good time to review ratios.
4. When students have finished their calculations, discuss their findings. Questions such as the following can be helpful:
   - What did you notice when you first started comparing tree seeds to mature trees?
   - How has actually calculating the ratio of seed mass to mature-tree mass affected your thinking?
5. Ask students to do a quick write in which they respond to the prompt, "What did you learn about trees and their seeds that you didn't know before?"
6. Wrap up this phase by introducing the question, "Where does the mass of a tree come from?"

***Assess this phase:*** Assessment in this phase is anecdotal and formative. Listen carefully to the conversations students have in small- and large-group discussions.

## Explore

In the Explore phase, students investigate where the mass of plants comes from by carefully measuring the mass of the soil and water used to grow bean seeds. Using texts to learn about photosynthesis, they identify carbon dioxide as another factor to be investigated. Students also gather evidence that plants take in carbon dioxide.

To accurately account for the mass of the water present in the soil and the fresh plant, you must dry samples of the soil and plants. This is a simple process that involves an oven and repeated measures of the soil and plants.

Dry the soil by heating a sample in the oven or microwave for a short period of time (200°F for 15 minutes in the oven or 20 seconds on high in a 900 W microwave). Measure the mass of the container and then the mass of the soil and container before heating and again after heating. Repeat the process until the mass is the same for two consecutive measurements. At this point, the water has been removed and only soil remains. This is the dry mass of the soil. Calculate the percentage of water in the original soil sample by dividing the dry mass of the soil (ending mass) by the wet mass of the soil (beginning mass) and multiplying by 100%. An example is shown in Figure 8.2.

### FIGURE 8.2. SAMPLE CALCULATION OF THE PERCENTAGE OF WATER IN A SOIL SAMPLE

$$\frac{\text{Dry mass of soil}}{\text{Wet mass of soil}} = \frac{15\text{ g}}{50\text{ g}} = 0.30 \times 100\% = 30\% \text{ (percentage of water in wet soil)}$$

Dry the plants in the oven in the same way you dried the soil. Students will have measured the mass of the plant and container in class. Put the plant and container in the oven at 200°F for 15 minutes. Remove the plant and container from the oven and calculate the mass again. Repeat the process until the mass is the same for two consecutive measurements. The plants will be very dry and brittle after all of the water has been removed.

Cupcake papers work well for the containers. Putting them in cupcake pans makes it easy to move them into and out of the oven. Because a low temperature is being used, the papers become warm, but do not burn. The temperature is well below the ignition point of the paper.

### PART I: GROWING THE PLANTS

Be sure to consider the following for Part 1 of the Explore phase:

- To allow time for the plants to grow, begin Part I of the investigation three to four weeks before you are ready to determine where the mass of a plant comes from.
- Teach a topic other than photosynthesis while the plants are growing, allowing a few minutes once or twice a week to water the plants.
- Dry a sample of the soil students will use for the investigation. This information is needed to fully calculate the mass of the water.

1. Open this phase by asking students, "What do you think a tree might need to grow?" Student responses are likely to include soil, water, sunshine, minerals, air, and so on.
2. Follow this with a discussion about which of these needs can be easily measured with the equipment available in the classroom. Guide the discussion toward soil and water. Inform students that they are going to plant some bean seeds, carefully measuring the soil and water to determine how each of these factors affects plant mass.
3. Organize the class into groups of three to four students. Provide each group with bean seeds, soil, five cups (one for each bean seed), water, a balance, and a graduated cylinder. Distribute a "Plant Mass Investigation and Data Sheet" to each student. Students will plant their seeds as directed.
4. Provide students with information on the percentage of water in the soil. Explain how the soil was dried. Students will use this information to determine the mass of the dry soil in each cup. The importance of knowing the mass of the dry soil will become apparent to students later in the investigation. At this point, they may not see the relevance of this information.
5. For several weeks, have students water their plants and record the volume of water used each time. Stop watering the plant approximately one week before beginning Part 2. It is easier to separate the plant from dried soil.

### PART II: MEASURING THE EFFECT OF SOIL ON PLANT MASS

1. Approximately one week after the last watering, measure the mass of the cup, plant, and soil. Then, have students carefully separate three of the plants from the soil. Save some plants for Part III. Students should separate each plant as follows:
    a. Remove the plant and soil from the cup. Work over a piece of paper towel or newspaper so that the soil is contained.
    b. Loosen the soil from the roots by hand, being careful not to break off pieces of the roots. Then gently shake the plant to remove additional soil.
    c. Use a small paint brush to remove as much of the remaining soil as possible.
    d. Fill a beaker or cup with water to about ¾ full. Swish the roots in the water to remove additional soil.
    e. Filter the water to remove the soil. Mass the filter paper before beginning. After filtering is complete, carefully remove the filter paper from the funnel, unfold it, and leave it overnight to dry. (*Note:* Some soil is likely to remain in the roots and some small root sections are likely to remain in the soil. The soil remaining is typically an insignificant amount and will not affect the outcome.)
2. Label the plants and the soil and let them air dry overnight.
3. The next day, measure the mass of each plant and its soil. Although they air dried overnight, a considerable amount of water is still in the plant and soil. Each group will put a massed sample of their soil in one plastic bag and a massed plant in another. Bags should be clearly labeled with the group name and mass of the contents.
4. Discuss the results with the class. Focus the conversation on the difference between the total amount of water used for the plant and the plant's mass. It should quickly become apparent that the amount of water used is greater than the increase in the plant's mass. Students are likely to note that some of the water is in the soil and some likely evaporated. They may draw a tentative conclusion that the mass of the plant comes from the water. Guide the discussion toward the need to dry the plant to see what the mass is without the water.
5. Dry the soil and plants as described on pages 142–143.
6. Now, have students determine the mass of the dried samples and compare the dry masses to the original masses of the soil and seeds. Students should find that changes in the mass of the soil are very minimal—considerably less than the gain in the plant mass. They should also find that the dry mass of the plant is greater than the mass of the seed, but considerably less than the mass of the water used.

Most of the water will have evaporated from the soil and transpired through the plant as it grew.

7. Lead a discussion about how the soil and water affected the mass of the plant. End with the question, "If the mass of a plant does not come from the water or the soil, where does it come from?" These questions may be helpful in guiding the students' thinking:

    - How does the mass of the plants compare to the average mass of the seeds?
    - How does the mass of the dry plant compare to the average mass of the seeds? What percentage of the plant was water?
    - How does the beginning mass of the dry soil compare to the ending mass of the dry soil? What does this tell you about how the soil affects the mass of the plant?
    - How does the mass of the water provided to the plants compare to the mass of the water in the plants and the soil? What does this tell you about how the water affects the mass of the plant?
    - According to what we know at this time, can you say conclusively whether the mass of the plant comes from the soil or the water? What additional information do you need?

8. Ask students to do a quick write in which they reflect on how the results of this investigation compare to what they thought would happen.

### PART III: MASS FROM THE AIR

At this point, students know that water contributes to the fresh mass of a plant and that some component other than water and soil is important to plant growth. It is now time to turn to texts for some answers. Students will identify carbon dioxide as another component to test and be introduced to photosynthesis.

The texts selected for this portion of the unit introduce students to the fundamentals of photosynthesis, without being overly technical. Information to be gleaned from this reading task includes the following:

- Photosynthesis is the process that plants use to make glucose.
- Plants use the glucose for food and thereby growth.
- Water, carbon dioxide, and sunlight are needed for photosynthesis.

A variety of texts, including online resources, are included to accommodate a wide range of reading abilities and access to materials (see pp. 135–138).

1. Provide students with a variety of texts (print and online) to read for this portion of the exploration. Instruct students to independently select two or more

resources to read. Students then read the resources, using the "Photosynthesis: Powering the Planet" graphic organizer to guide their thinking.

2. After students have completed the reading, discuss what they have learned.
3. Follow the discussion with an introduction to the testing for carbon dioxide investigation. Stress that because carbon dioxide is an odorless and colorless gas, they will need to use an indicator to test for its presence. Suggested discussion questions include the following:
   - What is the most interesting thing you learned from your reading?
   - Where did the raw materials—photosynthesis inputs—come from? Were you surprised by any of this?
   - Where do the products—photosynthesis outputs—end up?
   - How do plants use the glucose they produce?
4. Students will work in groups for the carbon dioxide test. Provide each group with a jumbo resealable bag, 150–200 ml of 4% BTB solution in a beaker or cup, a straw with a small cut near the bottom, and a test tube and stopper. Give each student a copy of the "Mass From the Air!" data sheet. Students will perform the investigation as described on the data sheet.
   a. Demonstrate to the class how to set up the experiment. Model blowing through the straw into the BTB solution. Caution students to blow into the straw instead of drawing the liquid up into the straw.
   b. Consider using a plant with lots of leaves when demonstrating how to set up the experiment. After demonstrating how to set it up, leave the plant in place overnight. The results are more dramatic for a plant with lots of leaves.
5. Students will check the results the following day. The control BTB should still be yellow. The BTB in the experimental setup will be blue or green.
6. Discuss the results.
7. Ask students to do a quick write in which they link this new information with information learned previously.

***Assess this phase:*** Assessment in the Explore phase includes content knowledge and science and literacy skills. The content knowledge is formatively assessed. When this phase is complete, students should know that (1) soil contributes very, very little to the mass of a plant, as shown by the minimal change in the beginning and ending mass of the soil; (2) more water is used in the growth of a plant than is retained by the plant, as shown by the total mass of the water given to the plant and the fresh mass of the plant; and (3) the mass of the dry plant comes from the carbon dioxide in the air, as shown by the BTB.

"Photosynthesis: Powering the Planet"; "Determining the Mass of a Tree"; "Plant Mass Investigation and Data Sheet"; "Mass From the Air!"; and the quick writes are all formative assessments for content knowledge. The science and literacy skills students use include reading and following procedural information, performing calculations using the data they have collected, conducting simple laboratory procedures (e.g., properly use a balance), and drawing conclusions from evidence. They may be formative or summative assessments for science and literacy skills, depending on whether the class has been working on developing specific skills.

## Explain

In this phase, students will respond to writing prompts to share what they know about where the mass of a tree comes from and how the glucose produced in photosynthesis is used.

1. Distribute a copy of "Speaking of Photosynthesis ..." to each student. Allow students sufficient time to record their evidence and respond to the prompts. Students will need to access their lab notes in order to include evidence from their investigations in their response.
2. As students work, circulate around the room and ask probing questions to individual students to gauge student understanding. If you see a common misconception cropping up as you talk with students, ask students to pause their work and discuss their thinking as a group.
3. Collect student work, review it, and either return it with suggestions for modifications (in the form of guiding questions) or discuss students' thinking as a class.
4. After students have reviewed the suggestions for modifications and participated in the class discussion, ask them to make any modifications they would like *in a different color of ink* (this will help you distinguish between initial and new work) *and* return it to you for evaluation.

***Assess this phase:*** Assessment in this phase is summative. The "Speaking of Photosynthesis ..." task gives students the opportunity to share what they know about photosynthesis and should be considered a summative assessment.

## Expand

In the Expand phase, students will investigate the effect of various wavelengths of light on photosynthesis. Students will design their experimental setups on the basis of what they learned in the Explore phase of the unit.

1. Review the importance of light in photosynthesis. Include information about the visible light spectrum that appeared in the reading materials students referenced in the Explore phase.
2. Show students a stoppered test tube containing a sprig of *Elodea* or *Cabomba* in a 1% sodium carbonate solution. Direct students to watch to see what is happening at the cut end of the plant.
3. After students have observed the bubbles coming off the cut end of the plant, ask them what they think the gas might be. Remind them of the products of photosynthesis and guide them toward oxygen.
4. Demonstrate how the test tube was setup and challenge them to work in teams to design an experiment to test how different wavelengths of light affect photosynthesis. Remind them to include a control.
5. Make different colors of cellophane, healthy sprigs of *Elodea* or *Cabomba,* test tubes, test tube racks, tape, grow lights, stopwatches, and scissors available to groups of students. You may want to include additional materials to encourage diverse thinking about the experimental design.
6. As groups are designing their experiments, circulate around the room and ask questions and check the experimental design. Questions could include the following:
   - What data will you collect?
   - How will you represent your data?
   - How could you represent your data in a graph?
   - Why are you setting up the experiment in this way?
   - What question are you investigating?
   - What do you hope to learn?
   - What role will each group member play?
7. After students have completed their experiments, ask each group to present their findings. Look for similarities across the experimental findings and use the information to draw a conclusion about how various wavelengths of light affect photosynthesis.

QR CODE 13

***Note:*** In lieu of the experiment, you may want to consider the virtual lab "Light and Plant Growth" *(http://bit.ly/1bePtGt;* QR code 13*)*. This website requires Adobe Flash, so the QR code may not open properly on mobile devices.

*Assess this phase:* The activities in this phase present a good opportunity to summatively assess students' science process skills. Use the Science Process Skills rubric in Appendix 1.

## REFERENCES

Canal, P. 1999. Photosynthesis and "inverse respiration" in plants: An inevitable misconception? *International Journal of Science Education* 21 (4): 363–371.

Driver, R. 1994. *Making sense of secondary science: Research into children's ideas.* London: Routledge.

National Governors Association Center for Best Practices and Council of Chief State School Officers (NGAC and CCSSO). 2010. *Common core state standards.* Washington, DC: NGAC and CCSSO.

NGSS Lead States. 2013. *Next Generation Science Standards: For states, by states.* Washington, DC: National Academies Press. *www.nextgenscience.org/next-generation-science-standards.*

Ozay, E., and H. Oztas. 2003. Secondary students' interpretation of photosynthesis and plant nutrition. *Journal of Biological Education (Society of Biology)* 37 (2): 68–70.

Name______________________________ Date____________________

# Comparing Masses

## CALCULATING THE AVERAGE MASS OF A TREE SEED

1. Mass each seed separately. Record each mass in a data table like the one below.
2. Add the masses together. Then, divide the sum by the total amount of seeds.

| Seed | Seed Mass |
|---|---|
| Seed 1 | 3.5 g |
| Seed 2 | 3.7 g |
| Seed 3 | 3.4 g |
| Total mass of seeds | 3.5 g + 3.7 g + 3.4 g = 10.6 g |
| Average mass of seeds | 10.6 g/3 = 3.53 g |

$$\frac{\text{Total mass of seeds}}{\text{Total number of seeds}} = \text{Average mass of a seed}$$

3. Record your seed data and calculate the average mass of a seed in the space below.

Name______________________________ Date______________

Comparing Masses (*continued*)

## DETERMINING THE MASS OF A TREE

1. Use a meter stick to measure to a spot on the tree that is 137 cm above the ground. Mark the spot with chalk.
2. Wrap a piece of string around the tree at the marked spot, and cut the string to the exact length of the circumference of the tree. Then, measure the string to determine the circumference of the tree in centimeters.
3. Find the circumference on the Mass of the Trunk of Hardwood Trees chart. What is your tree's mass in kilograms?
4. To find the ratio of seed to tree mass, divide the mass of the tree by the average mass of the seed. Remember to convert grams to kilograms (or kilograms to grams).

Example: Tree's circumference = 104 cm
Tree's mass = 835 kg
Average mass of seed = 3.53 g

Convert grams to kilograms: 3.53g/1,000 g = 0.00353 kg

$$\frac{\text{Mass of tree}}{\text{Mass of seed}} = \frac{835\text{ kg}}{0.00353\text{ kg}} = 236{,}543.9$$

In the example, the mass of the tree is 236,543.9 times as great as the mass of the seed!

Record your data and do your calculations.

### MASS OF THE TRUNK OF HARDWOOD TREES

| Circumference (cm) | Mass (kg) |
|---|---|
| 96.5 | 680 |
| 104 | 835 |
| 112 | 989 |
| 119 | 1,161 |
| 127 | 1,343 |
| 135 | 1,533 |
| 145 | 1,742 |
| 152 | 1,960 |
| 160 | 2,186 |
| 168 | 2,431 |
| 175 | 2,676 |
| 183 | 2,948 |
| 191 | 3,221 |
| 201 | 3,510 |
| 208 | 3,810 |
| 216 | 4,119 |
| 224 | 4,445 |
| 231 | 4,772 |
| 239 | 5,126 |
| 246 | 5,479 |
| 254 | 5,851 |
| 264 | 6,232 |
| 272 | 6,622 |
| 279 | 7,030 |
| 287 | 7,448 |

*Source:* Data from "Landowner's Guide to Determining Weight of Standing Hardwood Trees," n.d., by David W. Patterson and Paul F. Doruska, Division of Agriculture, University of Arkansas, *http://bit.ly/1PjpI9L.*

Name________________________________ Date________________

# Plant Mass Investigation and Data Sheet

## PART I: PLANTING THE SEEDS

### Materials

- 6 cups for planting
- 5 bean seeds
- 50 ml graduated cylinder
- Balance
- Soil
- Water
- Marker
- Goggles
- Gloves
- Aprons

### Safety Reminders

- Wear goggles, gloves, and aprons during all phases of the investigation, including during setup and cleanup.
- Be careful when working with glass or plastic containers.
- Do not eat the beans.
- Wash your hands with soap and water after completing the investigation.

### Background

When plants begin growing, they use the food that is stored in the seed. By the time they sprout and the first leaves appear, all of the stored food from the seed has been used. Yet plants continue to grow. How can that be? You're about to find out by planting a few bean seeds, carefully measuring the soil, seeds, water (1 ml of water has a mass of 1 g), and eventually the plant and by conducting some simple experiments.

1. Determine the average mass of the seeds. Record your data and calculations in Table 1.
2. Label each cup with the date and team name. Number the cups 1–5 and label one "control."
3. Mass each cup. Record the mass for each cup in Table 2.
4. Fill each cup approximately ¾ full with soil. Mass the cup and soil. Record the data in Table 2.
5. Add a single bean seed to cups 1–5. Do not add a seed to the control cup. Push the seed about ½ inch into the soil. Cover the seed with soil. Gently press down on the soil that is over the seed.

Name________________________________________ Date_______________________

### Plant Mass Investigation and Data Sheet (*continued*)

6. Put 50 ml of water into the graduated cylinder. Slowly add water to one of the cups until the soil is completely moistened but not too wet. There should not be puddles of water on top of the soil. You will not use all of the water. You may have to add water, pause, and then add more water. Record the volume of water added to the cup in Table 3. Now, add the same amount of water to each cup.
7. Set the cups on a sunny windowsill or under a grow light.
8. Water the seeds/plants twice a week for three to four weeks. Record the volume of water added to each cup.
9. Stop watering the plants one week before beginning Part 2 of the investigation. Calculate the total mass of the water added to the plants.

## PART II: MEASURING CHANGES IN SOIL AND WATER MASS

### Materials

- Balance
- Parchment paper labels
- Markers
- Zipper bags (lunch size)
- Large sheets of paper or plastic to cover the work area
- Beakers or cups
- Water
- Goggles
- Gloves
- Aprons

### Safety Reminders

- Wear goggles, gloves, and aprons during all phases of the investigation, including during setup and cleanup.
- Be careful when working with glass or plastic containers.
- Do not eat the beans.
- Wash your hands with soap and water after completing the investigation.

### Day 1

1. Cover your work area with plastic or paper. Carefully remove three plants and the soil from their cups. Make sure all of the soil is removed. Take care not to drop soil on the floor or outside of the paper so that measurements of the mass of the soil are accurate.
2. Hold the stem of the plant near the soil and gently shake the plant to remove the soil. Some soil will remain on the roots. Remove the remaining soil by hand with a soft-bristle brush or by swishing the roots in a beaker filled with water.

Name________________________________________ Date_______________________

## Plant Mass Investigation and Data Sheet *(continued)*

a. Soil that was removed by swishing the roots in water will have to be dried before it is massed. Mass the filter before beginning. Filter the water to remove the soil. Set the filter aside overnight to dry.

b. Mass a piece of paper towel. Record the mass. Spread the remaining soil out on the paper towel to dry overnight.

3. Set the soil and plants aside overnight to dry.
4. Water the two remaining plants.

### Day 2

1. Mass each plant and record the data.
2. Use the parchment paper strips to label each plant with your team's name and the plant number.
3. Mass all the soil samples and determine the total mass.
4. Discuss the following questions with your group.
    - How does the mass of the soil compare to the original mass?
    - How does the mass of the plant compare to the mass of the seed?
    - Could the mass of the plant have come from the soil? Explain your answer.
    - How does the mass of the plants compare to the mass of the water provided to them?
    - Could the mass of the plant have come from the water? Explain your answer.
5. Discuss with your group possible explanations for the differences between the mass of the water and the mass of the plants.

Name______________________________ Date________________

## Plant Mass Investigation and Data Sheet (*continued*)

Calculate the average mass of the seeds. Record your data in Table 1, and show your calculations in the space provided in the table.

### TABLE 1. AVERAGE MASS OF THE SEEDS

| Data | Calculations |
|---|---|
| | Average mass of the seeds |

Calcuate the average mass of the cup, and record your data in Table 2. Use the following equations to calculate the beginning mass of the soil and cup:

- Mass soil – mass cup = mass wet soil
- Wet soil × % water in dry sample = mass of water in soil
- Wet soil – mass of water in soil = mass of dry soil

### TABLE 2. SOIL DATA: AVERAGE MASS OF CUP

| Data Information | Example | Cup 1 | Cup 2 | Cup 3 | Cup 4 | Cup 5 | Control |
|---|---|---|---|---|---|---|---|
| Mass of cup | 5 g | | | | | | |
| Mass of cup and soil | 55 g | | | | | | |

Time for calculations! Find the mass of the wet soil by subtracting the mass of the cup from the mass of the cup and soil.

Example: 55 g (Mass of cup and soil) – 5 g (Mass of cup) = 50 g (Mass of wet soil)

Name______________________________ Date______________

## Plant Mass Investigation and Data Sheet (*continued*)

| Data to Determine | Example | Cup 1 | Cup 2 | Cup 3 | Cup 4 | Cup 5 | Control |
|---|---|---|---|---|---|---|---|
| Mass of wet soil | 50 g | | | | | | |
| Mass of water in soil | 15 g | | | | | | |

Time for calculations! Find the mass of water in the soil by multiplying the mass of wet soil by the percentage of water in the dry sample (your teacher has this data).

Example:

40 g × 0.30 = 12 g

| 50 g | × | 0.30 | = | 15 g |
|---|---|---|---|---|
| Mass of wet soil | | Percentage of water in wet soil | | Mass of water |

| Data to Determine | Example | Cup 1 | Cup 2 | Cup 3 | Cup 4 | Cup 5 | Control |
|---|---|---|---|---|---|---|---|
| Mass of dry soil | 35 g | | | | | | |

Time for calculations! To determine the mass of the dry soil, subtract the mass of the water from the mass of the wet soil.

Example:

40 g wet soil – 12 g water = 28 g dry soil

| 50 g | – | 15 g | = | 35 g |
|---|---|---|---|---|
| Mass of wet soil | | Mass of water in wet soil | | Mass of dry soil |

Name______________________________ Date____________________

Plant Mass Investigation and Data Sheet (*continued*)

**TABLE 3. WATER DATA**

| Date | Volume of Water | Mass of Water (1ml of water = 1 g) |
|---|---|---|
| | | |
| | | |
| | | |
| | | |
| | | |
| | | |
| Total | | |

Name______________________________ Date__________________

# Photosynthesis: Powering the Planet

This picture tells the story of photosynthesis, but the story is incomplete. Fill in the boxes with the resources and products of photosynthesis to complete the story. Diagrams in the texts you chose will provide helpful information.

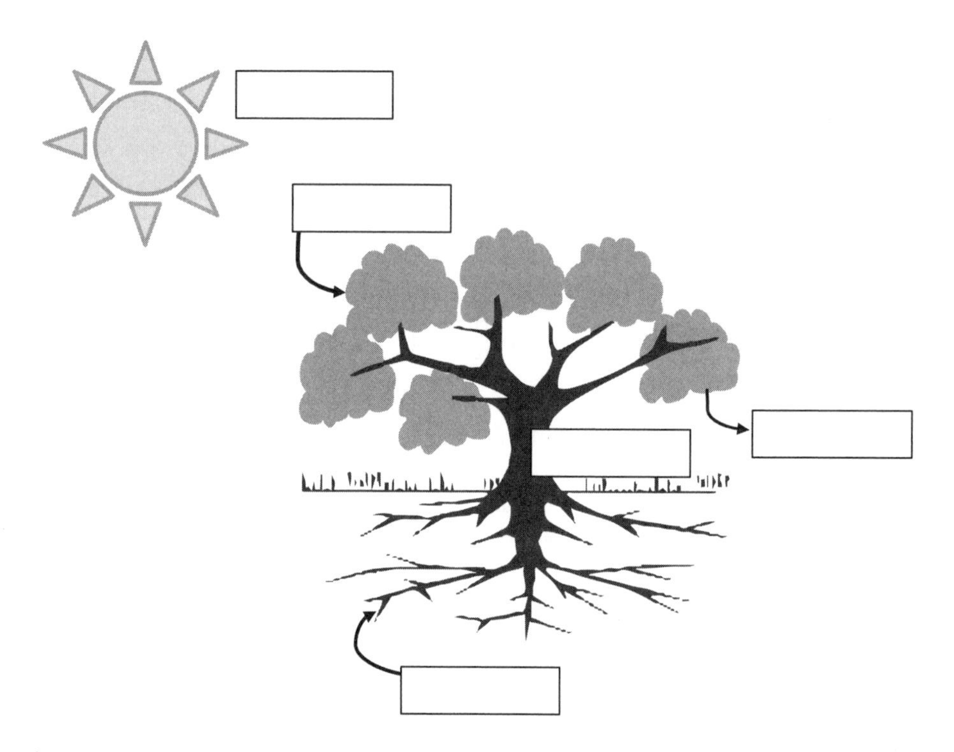

As you read, make notes about what the plant takes in (photosynthesis inputs) and what the plant makes or releases (photosynthesis outputs).

| Photosynthesis inputs | Photosynthesis outputs |
| --- | --- |
| | |

How do plants use the output products of photosynthesis? Write your answer in the space below.

Name________________________________________ Date_______________________

# Mass From the Air!

## MATERIALS

- Bromothymol blue (BTB) solution (250 ml)
- Beaker or cup
- Straw with a small hole in the bottom (1 per student)
- Test tube and stopper or a vial with a cap
- Large zipper bag (2-gallon size)
- Bean plant
- Small container
- Goggles
- Gloves
- Aprons

## SAFETY REMINDERS

- Wear goggles, gloves, and aprons during all phases of the investigation, including during setup and cleanup.
- Be careful when working with glass or plastic containers.
- Follow your teacher's instructions in using the BTB solution safely.
- Wash your hands with soap and water after completing the investigation.

## BACKGROUND

BTB is an acid indicator that is used to tell the difference between an acid and a base. It is bluish-green when it is neutral. BTB turns yellow in an acid and blue in a base. You can change BTB from bluish-green (neutral) to yellow by using a straw to blow the carbon dioxide ($CO_2$) you exhale into the solution. The $CO_2$ combines with the water in the solution to make a weak carbonic acid solution, thus turning the indicator yellow. When the $CO_2$ that has been added to the solution is removed, the solution returns to a bluish-green color.

## PROCEDURE

1. Label the large zipper bag with your team's name and the date. Put one or two of your bean plants in the bag. Set the bag aside.
2. Using the straw with a small hole in the end of it, gently blow into the BTB solution. (The end with the small hole goes in the solution.) If you are taking turns blowing to the solution, each student must use his or her own straw. Stop blowing into the BTB when the solution turns yellow.
3. Label the control test tube with your team name. Fill the test tube with the yellow BTB solution. Put the stopper in the test tube and set it aside.

Name________________________________ Date__________________

## Mass From the Air! (*continued*)

4. Pour a small quantity of the yellow BTB into the small container. Place the container in the bag that has the plants in it. Seal the bag and set it under the grow light or on a sunny windowsill.
5. The next day, compare the control solution with the solution in the bag with the plants. Record your results.

| Color of control BTB ______________ | Color of test BTB ______________ |
|---|---|

What does your evidence indicate about what happened to the carbon dioxide in the BTB solution? Where might the carbon dioxide have gone? Record your answers in the space below.

Name________________________ Date________________

# Speaking of Photosynthesis ...

Tyla and Audrey have stopped for a bite to eat after school. Owen hears them talking about their science assignment and stops to join the conversation. They are trying to answer the question, Where does the mass of a tree come from? **Do you agree with Tyla, Audrey, or Owen?**

In the space below, record evidence from class and reading that will help you make a decision. Then, write the name of the character you agree with and your reasons for agreeing with him or her in the bottom space.

| Evidence from class | Evidence from reading |
|---|---|
| | |

Do you agree with Tyla, Audrey, or Owen? Explain using the evidence from class and reading.

Name________________________________________ Date________________________

## Speaking of Photosynthesis ... (*continued*)

Audrey: What happens to the glucose the plants make?

Tyla: I guess animals eat it.

Tyla and Audrey are wondering where the glucose plants produce ends up. **How would you answer Audrey's question?** In the space below list evidence from your reading that will help you answer Audrey's question. Include the source of the evidence. Then, write a response to Audrey's question in the bottom space.

| Evidence | Source |
| --- | --- |
| | |

Response to Audrey's question

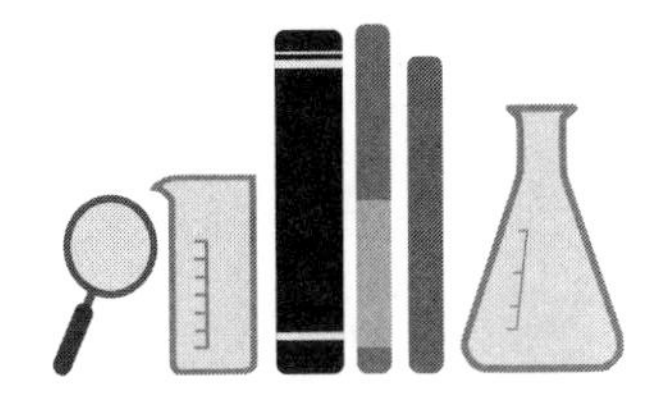

## Chapter 9

# Chemistry, Toys, and Accidental Inventions

## OVERVIEW

In this unit, students will demonstrate that a chemical change results in a new substance and that the properties of the new substance are substantially different from the beginning substances. Students read about the accidental invention of Silly Putty and then make their own putty from common household substances. Next, they conduct a series of chemical reactions in which they carefully collect data about the properties of the reactants and products. Students are able to confirm that the products have different properties than the reactants and that these properties can be quantified. Along the way, they discover that matter is neither created nor destroyed in a chemical reaction. Ultimately, students use the data they collect to answer the question, *Was there a chemical reaction?*

This unit assumes that students know and can demonstrate that matter is made up of atoms, that substances can be described by their properties, and that different substances have different properties. Students should be familiar with the concepts of density, solubility, and pH. They should be able to use basic lab equipment, including balances. Finally, it is helpful if students have some experience drawing evidence-based conclusions. If students are just developing the prerequisite knowledge and skills, plan time for mini-lessons to support students as they work through the unit.

## OBJECTIVES

- Distinguish between reactants and products.
- Explain the law of conservation of matter.
- Measure properties of reactants and products.
- Recognize that a change in the reactants results in a difference in the products.
- Observe that a chemical reaction results in a new substance with different properties than the original substance.
- Synthesize information from text and lab activities to define vocabulary.

## STANDARDS ALIGNMENT

### *Next Generation Science Standards* (NGSS Lead States 2013)

#### PS1.B: CHEMICAL REACTIONS

- Substances react chemically in characteristic ways. In a chemical process, the atoms that make up the original substances are regrouped into different molecules, and these new substances have different properties from those of the reactants. (MS-PS1-2, MS-PS1-3,MS-PS1-5)

- The total number of each type of atom is conserved, and thus the mass does not change. (MS-PS1-5)

### *Common Core State Standards, English Language Arts* (NGAC and CCSSO 2010)

#### GRADES 6–12 LITERACY IN HISTORY/SOCIAL STUDIES, SCIENCE, AND TECHNICAL SUBJECTS

- CCSS.ELA-LITERACY.RST.6-8.3: Follow precisely a multistep procedure when carrying out experiments, taking measurements, or performing technical tasks.
- CSS.ELA-LITERACY.RST.6-8.9: Compare and contrast the information gained from experiments, simulations, video, or multimedia sources with that gained from reading a text on the same topic.

## TIME FRAME

- Four to five 45-minute class periods

## SCIENTIFIC BACKGROUND INFORMATION

Chemical reactions are an important part of our day-to-day existence. We don't always recognize when we are performing chemical reactions because they are a routine part of what we do. All of these routine behaviors involve chemical reactions: driving a car, baking a cake, frying a hamburger, burning a candle, boiling an egg, washing dishes, doing laundry, and even taking a shower! When chemical reactions occur, the atoms in the starting substance(s) are rearranged and recombined into a new substance with completely different properties.

The substances we begin with in a chemical reaction are called the *reactants.* A chemical reaction can begin with just one reactant or with several. The substances we end up with after a chemical reaction are called the *products.* A chemical reaction can end with just one product or with several. For example, the formation of table salt begins with two reactants, sodium and chlorine, and ends with one, sodium chloride. The reaction between baking soda and vinegar begins with those two reactants and ends with three products—sodium acetate, carbon dioxide, and water.

Regardless of the number of reactants and products, a chemical reaction begins and ends with exactly the same number of atoms—the same mass. For example, if the reactants are 35.5 g of chlorine and 23 g of sodium, the product will be 58.5 g of sodium chloride. If the reactants are 84 g of baking soda and 60 g of vinegar, the products will be 18 g of water, 44 g of carbon dioxide, and 82 g of sodium acetate. This phenomena is known as the *law of conservation of mass (matter).* Matter can neither be created nor destroyed in a chemical reaction. The atoms in the reactant can be rearranged to form new products, but the amount of mass at the beginning is equal to the amount of mass at the end. Measuring and comparing the mass of the reactants and products provides evidence for this law.

In middle school science, we often cite several pieces of evidence that a chemical change has occurred: the evolution of heat, light, or a gas; formation of a precipitate; a change in temperature; or a change in color or odor. All of those are qualitative observations. They could be measured, but often they are not. Moreover, we could argue that some of those things can happen in a physical change.

We often do not spend enough time talking about the different properties of the reactants and products. A change in those properties is indisputable evidence that a chemical reaction has occurred. Analyzing and comparing the properties of the reactants to the properties of the products makes distinguishing between a physical and chemical reaction more exact. We can easily compare the density, solubility, and pH of the reactants and products in many chemical reactions. However, not all components of a reaction are visible; it's difficult to measure these properties in a gas, for example. But many reactants and products can be easily observed—certainly enough for middle and high school students to recognize that the product is qualitatively and quantifiably different from the reactants.

## MISCONCEPTIONS

Common misconceptions about chemical reactions are discussed in Table 9.1.

### TABLE 9.1. COMMON MISCONCEPTIONS ABOUT CHEMICAL REACTIONS

| Common Misconception | Scientifically Accurate Concept |
| --- | --- |
| Dissolving is a chemical change. | Dissolving is a physical change in which the particles of the solute are evenly distributed in the solvent. The properties of the solute and the solvent are not changed. |
| Changes in phase are chemical changes. | No new substance is formed in a phase change. When ice melts into water and then evaporates to become water vapor, the chemical structure of the molecules does not change. All of the water molecules are still $H_2O$. |
| Mass is not conserved. | Even in the most extraordinary chemical reaction, mass is conserved. It is conserved at the system level. That means that when we are able to contain all of the reactants and products, nothing escapes into the environment. Therefore, the mass is unchanged. This misconception may linger because of everyday experiences, such as a campfire, when it appears that mass has changed. |
| • Chemical changes are additive.<br>• The original substances remain after a chemical reaction, even though they are altered. | The original substance has been changed into a new substance. The new substance has different properties than the original substances. |

*Source:* American Institute of Physics 2016; Barker 2004.

## MATERIALS

- Silly Putty
- White glue (see safety data sheet information in Appendix 4)
- Liquid starch (see safety data sheet information in Appendix 4)
- Tongue depressors
- Beaker or bowl for mixing putty
- pH strips
- Filter papers
- Funnels
- Silicone oil (see safety data sheet information in Appendix 4)
- Boric acid solution (see safety data sheet information in Appendix 4)
- Baking soda (see safety data sheet information in Appendix 4)
- Vinegar (see safety data sheet information in Appendix 4)
- Various fats used in soap making (e.g., coconut oil, olive oil, beeswax, vegetable shortening, canola oil)
- Borax (see safety data sheet information in Appendix 4)
- Castile soap (see safety data sheet information in Appendix 4)
- Laundry detergent (see safety data sheet information in Appendix 4)
- Baby powder
- Stirring rod or coffee stirrers
- Balances (1 per group)
- Graduated cylinders
- Beakers, bowls, or plastic cups for the chemical reactions
- Goggles (sanitized, indirectly vented chemical splash)
- Nonlatex gloves
- Aprons

## SUPPORTING DOCUMENTS

- "Comparing Properties of Reactants and Products" investigation sheet
- "Chemical Reaction: Frayer Model" diagram
- "Was There a Chemical Reaction?" graphic organizer
- "Gakish Recipe Cards"

## SAFETY CONSIDERATIONS

- Have students wear goggles, gloves, and aprons during all phases of the investigation, including during setup and cleanup.
- Review safety information in safety data sheets for hazardous chemicals with students.
- Remind students to use caution in working with glassware or plasticware, which can puncture skin if broken.
- Make sure students have appropriate procedures in place for disposal of excess reactants and products from the investigation.
- Have students wash their hands with soap and water after completing the investigation.

## NONFICTION TEXTS

QR CODE 1

"Basic Slow Cooker Soap" by Wellness Mama, 2016, *http://bit.ly/1KNQGU0.* (QR code 1)

This blog post discusses the basic ingredients and process of making soap at home.

QR CODE 2

"Lead Iodide Precipitating" by Hans de Grys, 2014, *http://bit.ly/1PVX02v.* (QR code 2)

This dramatic video demonstrates the chemical reaction between lead nitrate and sodium iodide. One of the products is lead iodide, which is a bright yellow compound that precipitates out of the solution.

QR CODE 3

"Reactants and Products" by CK–12 Foundation, 2016, *http://bit.ly/1ThzYEk.* (QR code 3)

This short article explains the difference between reactants and products in a chemical reaction. It goes on to explain the basics of a chemical reaction.

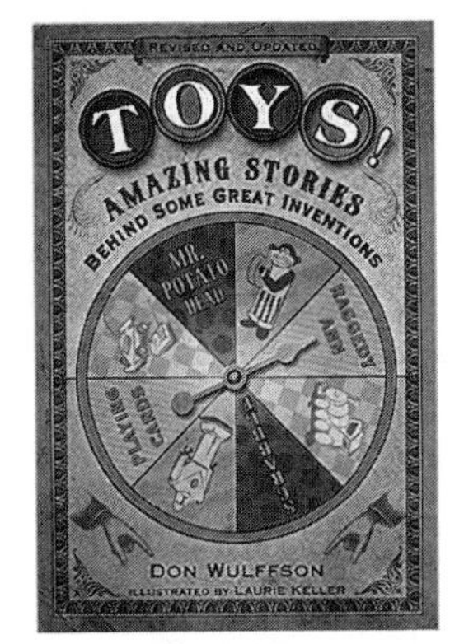


*Toys! Amazing Stories Behind Some Great Inventions* by Don Wulffson (New York: Square Fish, 2014); published Lexile level 920L.

This book shares the amazing stories of how some of the most popular toys ever made were the result of accidental inventions.

## LEARNING-CYCLE INQUIRY

### Engage

In this phase, students will learn how Silly Putty was accidentally invented and will make their own putty. They then conduct a series of chemical reactions in which they carefully collect data about the properties of the reactants and products. Ultimately, they are able to confirm that the products have different properties than the reactants and that these properties can be quantified.

1. Begin by passing around enough Silly Putty for students to play with and share. One package of Silly Putty for every three to four students should suffice. Let students play with the Silly Putty for about five minutes. Talk about whether they have ever seen or played with Silly Putty before. Ask students if they have ever thought about how it was invented.
2. Ask students to independently read about how Silly Putty was invented. Provide each student with a copy of "Silly Putty" from *Toys! Amazing Stories Behind Some Great Inventions.*
3. When students have finished reading, write the following sentence starters on the board:

   - I thought it was interesting that …
   - I was surprised by …
   - I am wondering …
   - I learned that …

   Ask students to select one of the four sentence starters and compose an answer. Students may do this mentally or as a quick write.
4. Assign each corner of the room (or four areas in the room) one of the sentence starters. Students should then move to that corner (area) and share their responses in these small groups.
5. Ask for one or two volunteers from each corner to share either their response or a response that they liked from the discussion.
6. Have students return to their seats. Ask students how many chemicals were combined to make Silly Putty (two). Show students samples of silicon oil and boric acid. Marvel aloud that these two chemicals could combine to make something as different as Silly Putty. Explain to students that they will now create their own putty using a homemade recipe and compare the finished product to the commercial product.
7. Make the putty.

a. Organize the class into groups of three to four students. Provide each group with a two-to-one ratio of white glue to liquid starch, a tongue depressor, and a beaker or bowl for mixing the glue and starch. (Depending on the commercial products you use, you may need to adjust the proportions. Test them in advance.)

b. Instruct students to observe the two liquids and list their characteristics.

c. After observing the liquids, pour them into the bowl and stir the solution with a tongue depressor. Continue stirring until the mixture is the same consistency as Silly Putty.

d. Remove the results. Knead and stretch the product until it feels like Silly Putty.

e. Play with the homemade putty, comparing its properties to the commercial version. After students have finished manipulating the homemade putty, they should draw a Venn diagram and compare their putty to Silly Putty.

f. After students have had sufficient time to compare, discuss the similarities between the two products.

*Assess this phase:* Assessment in this phase is formative in nature. Anecdotal evidence from discussions, students' answers to the four corners questions, and the Venn diagrams all provide a glimpse of what students are thinking at this time. This activity may reveal some student misconceptions.

## Explore

In this phase, students will continue working in groups to examine and compare the properties of the reactants and products of a chemical reaction. They will also keep track of the mass of the reactants and products. In the case of soap, they will not examine sodium hydroxide. It is extremely basic, and it can cause burns if it comes into contact with skin. Sodium hydroxide is also very dangerous if it splashes in someone's eyes. Thus, students will examine the other reactants and the product of soap making.

*Advance preparation:* Organize the reactants, products, and equipment in a way that will facilitate efficiency and ensure that students use the correct reactants for each chemical reaction.

After preparations are made, begin the Explore phase. The steps are as follows:

1. Distribute a copy of "Comparing Properties of Reactants and Products" to each student.
2. Allow groups to conduct the investigation as written, stopping when they get to the "Analyzing My Data" table. Depending on their lab experience, students may need you to model the following procedures:
   - Proper use of pH strips

- Water displacement
- Proper folding of filter paper to use with a funnel
- Manipulation of the nonlatex glove for the "Giant Hand" portion of the investigation. For the demonstration, use something that doesn't react with water (e.g., flour, salt, sugar) in place of the baking soda. Use water in place of the vinegar. That way, the inflation of the hand with use of carbon dioxide will be a surprise for students.
- Proper use of a balance, graduated cylinder, and other lab tools

3. As students are working through the investigation, walk around the room asking guiding questions and providing assistance where needed. Remember to keep a watchful eye on students to make sure they are following the safety procedures as directed.
4. Bring the class back together to discuss the results. Project the "Analyzing My Data" table where all students can see it. Model how to complete the table. Students may need support as they work with their data. After students have had a chance to complete the table and discuss it with their group, bring the class back together to discuss their results. At the end of the discussion, students should know that after the chemical reaction takes place, a new substance that is different from the original substances is formed.
5. Students should now finish their investigation. When they are finished, discuss the law of conservation of mass. On the basis of their evidence and the class discussion, students should recognize that the mass of the reactants and the mass of the products are the same. Ask the students to do a quick write in which they respond to the question, "How do the mass of the reactants compare to the mass of the products in a chemical reaction?"
6. Distribute the "Chemical Reaction: Frayer Model" diagram to students. Ask students to independently complete the diagram. After they are finished, they should share and discuss their work with a classmate. Follow this step with a class discussion.

*Assess this phase:* The Explore phase offers opportunities to formatively and summatively assess students. The individual, small-group, and large-group discussions can be used as formative assessments. The Frayer model can serve as a summative assessment. If students do not demonstrate proficiency on the Frayer model, consider revisiting the parts of the Explore phase that will best meet their needs.

### Explain

In this phase, students will share what they have learned about chemical reactions and the conservation of matter. Students will view a video ("Lead Iodide Precipitating") of a chemical reaction between lead nitrate solution and sodium iodide solution. In the video, the two clear colorless solutions are combined to produce a bright yellow precipitate, lead iodide. Provide each student with the "Was There a Chemical Reaction?" graphic organizer.

1. Give students time to complete the graphic organizer.
2. Provide support where needed.

*Assess this phase:* Use students' responses to the prompt as a summative assessment.

### Expand

In this phase, students will compare the results of different "recipes" for making Gak-like substances (Gakish) with white glue.

1. Give students the recipe cards for two additional forms of Gak.
2. Combine 60 ml of white glue with 35 ml of laundry detergent.
3. Combine 5 tbs of baby powder with with 20 ml of white glue.
4. As students follow the chemical reaction recipes, they should collect and analyze data as they did in the Explore phase.
5. Students will now compare these products with the putty and Gak made earlier.
6. Ask students to write an advertisement for the product they like best. Include three reasons why their choice is better than the others.

*Assess this phase:* Students' advertisements serve as an additional form of summative assessment. Their work should be assessed to determine their understanding of chemical change.

## REFERENCES

American Institute of Physics. 2016 Children's misconceptions about science. *http://amasci.com/miscon/opphys.html.*

Barker, V. 2004. Beyond appearances: Students' misconceptions about basic chemical ideas. Report prepared for the Royal Society of Chemistry, London. *http://modeling.asu.edu/modeling/KindVanessaBarkerchem.pdf.*

National Governors Association Center for Best Practices and Council of Chief State School Officers (NGAC and CCSSO). 2010. *Common core state standards.* Washington, DC: NGAC and CCSSO.

NGSS Lead States. 2013. *Next Generation Science Standards: For states, by states.* Washington, DC: National Academies Press. *www.nextgenscience.org/next-generation-science-standards.*

Name________________________________________ Date____________________

# Comparing Properties of Reactants and Products

For each of the following chemical reactions, you will be comparing the properties of the starting substances (reactants) and ending substances (products). Carefully read and follow the directions as you perform the chemical reactions. You must wear goggles and an apron (or lab coat).

Before beginning, read this short article online: "Reactants and Products" (*http://bit.ly/1ThzYEk;* see QR code). The article provides some background information that will be helpful.

QR CODE: Reactants and Products

## PROCEDURE

1. Collect the following information for each of the products and reactants:
   a. Density
      i. Liquids: Mass 15 ml of the liquid. Divide the mass by 15 ml to determine the density.

| Substance | Mass (g) | Volume | Density (mass/volume = g/ml) |
|---|---|---|---|
| Oil #1 for Soap | | 15 ml | |
| Oil #2 for Soap | | 15 ml | |
| White glue | | 15 ml | |
| Borax solution | | 15 ml | |

      ii. Soap: Measure the height, width, and length of the soap sample. Multiply them to determine the volume. Mass the bar of soap. Divide the mass of the soap by its volume.

| Substance | Mass (g) | Volume (height × width × length = $cm^3$) | Density (mass/ volume = $g/cm^3$) |
|---|---|---|---|
| Soap | | | |
| Oil for soap | | | |

      iii. Semisolid fats and Gak: Measure the mass of the sample. Use water displacement to measure the volume.

Name________________________________ Date____________________

## Comparing Properties of Reactants and Products (*continued*)

| Substance | Mass (g) | Volume (ml) | Density (mass/ volume = g/ml) |
|---|---|---|---|
| Semisolid fat | | | |
| Gak | | | |

b. Solubility

i. Liquids and semisolid fat: Pour a small quantity of the liquid or small piece of semisolid fat reactant into room temperature water. Stir gently for about 30 seconds. If it is soluble, a heterogeneous (all parts are alike) mixture will result. If it is insoluble, it will float on top of the water or sink.

ii. Solids: Put about 75 ml of warm water in a beaker or cup. Use a different cup for each test. Label the cups with the name of the reactant you will be testing. Add a few grams of the solid to the labeled cup of warm water. Stir for about 30 seconds. The solid will dissolve into the solution and will no longer be visible. For the soap, shave off small thin pieces to test the solubility. If the solid dissolves, save some of the water for pH testing.

c. pH

i. Liquids: Dip a stirring rod or a coffee stirrer into the liquid. Touch the rod to a pH strip. Compare the color of the pH strip to the pH scale to determine the pH of the liquid.

ii. Solids: Use the water from the solubility test to conduct the pH test for the solid in the same way the liquids were tested.

iii. Semisolid fat: Use a stirring rod to put a very small sample of the semisolid fat on a pH strip.

d. Mass

i. Liquids: Measure the mass of the graduate cylinder to be used. Pour in the correct amount of liquid. Mass the liquid and graduated cylinder together. Determine the mass of the liquid by subtracting the mass of the graduated cylinder.

ii. Solids and semisolids (reactants): Determine the mass of a weighing boat or weighing paper. Add the required amount of the reactant (if the weighing boat has a mass of 1.0 g and you need 15 g of the reactant, the final reading on the balance should be 16 g).

iii. Solids and semisolids (products): Determine the mass of a weighing boat or weighing paper. Put *all* of the solid or semisolid product in the weighing boat. Record the mass. Then, subtract the mass of the weighing boat to find the mass of the product.

Name________________________________ Date____________________

## Comparing Properties of Reactants and Products (*continued*)

# SOAP

Soap is produced by combining sodium hydroxide (lye) with some kind of fat. The fats used can be oils of solids such as vegetable shortening. The kind and amount of fat used determines the properties of the soap. The kind of oil used can determine how hard or soft the soap is, how much lather it makes, how well it cleans, and how well it moisturizes and conditions skin. Because sodium hydroxide is highly corrosive and can cause serious burns, the information has been provided for you.

Before beginning, read this short article online: "Basic Slow Cooker Soap Recipe" (*http://bit.ly/1KNQGU0*; see QR code). The article provides information about soap making. As you read, think about the properties of the basic ingredients used in soap making.

QR CODE: Basic Slow Cooker Soap

## The Reaction

1. Because of the corrosive nature of sodium hydroxide, you will not be making soap, even though many people do make their own soap.

| | Reactants | | | → | Product |
|---|---|---|---|---|---|
| | Sodium Hydroxide | + | Fats | → | Soap |
| pH | Strongly basic, pH = 11 – 14 | | | | |
| Soluble in Water | Yes | | | | |
| Density | 2.13 g/cm$^3$ | | | | |
| State of Matter | Solid | | | | |

# GAK

Gak is produced in the same way the putty is produced, but a borax solution is used in place of the liquid starch. Borax and liquid starch are very different chemically. Borax is a mineral that is mined from Earth. The borax used in this reaction is a borax and water solution. Starch is a substance produced by plants. The liquid starch used in this reaction is a solution of starch and water.

## The Reaction

1. Pour 50 ml of white glue into a beaker. Add 50 ml of water. Stir.
2. Pour 50 ml of water into a different beaker. Add 8.5 g of borax to 120 ml of water. Stir.

Name______________________________ Date____________________

## Comparing Properties of Reactants and Products (*continued*)

3. Slowly add the borax solution to the glue solution, stirring the whole time.
4. Remove the Gak from the cup. Knead the Gak for a few minutes until it is the desired consistency.

| | **Reactants** | | | → | **Product** |
|---|---|---|---|---|---|
| | White Glue | + | Borax Solution | → | Gak |
| Soluble in Water | | | | | |
| Density | | | | | |
| State of Matter | | | | | |
| Mass | | | | | |
| Total Mass | | | | | |

## BAKING SODA AND VINEGAR

The baking soda and vinegar chemical reaction is a well-known reaction that uses simple household items. The reaction is very quick and always fun to watch.

1. Put 5 g of baking soda in a medium-size beaker or cup.
2. Add 30 ml of vinegar.
3. Filter the water off the products.
    a. Mass a filter paper. Record the mass.
    b. Fold the filter paper into quarters as shown.
    c. Form a cone by pulling one layer of the filter paper away from the remaining three.
    d. Put the cone into the funnel. Set the funnel in a graduated cylinder.
    e. After the water stops dripping out the funnel, remove the funnel and take out the filter paper. Set the filter paper and sodium acetate aside to dry overnight.
    f. Measure the volume of water in the cylinder. The mass of 1 ml of water is 1 g. So the number of milliliters is equal to the number of grams of water in the products.

Name______________________ Date______________

## Comparing Properties of Reactants and Products (*continued*)

| | Reactants | | | → | | | Products | | |
|---|---|---|---|---|---|---|---|---|---|
| | Baking Soda | + | Vinegar | → | Sodium Acetate | + | Carbon Dioxide | + | Water |
| pH | | | | | | | | | |
| Soluble in Water | | | | | | | | | |
| Mass | | | | | | | | | |
| Total Mass | | | | | | | | | |

How do the products compare to the reactants in each chemical reaction? Use the table below to help organize your thinking. Some of the cells will not be used.

### ANALYZING MY DATA

| Data | Soap | Gak | Baking Soda and Vinegar |
|---|---|---|---|
| **Density**<br>How did the density change? (increase, decrease, stay the same) If you have two reactants and one product, compare the density of the product to each of the reactants. | | | |
| **Solubility**<br>Is the product more or less soluble than the reactants? If you have two reactants and one product, compare the solubility of the product to each of the reactants. | | | |
| **pH**<br>How did the pH change? Compare all of the reactants of all of the products. | | | |
| **Mass**<br>How did the total mass of the reactants compare to the total mass of the products? | | | |
| **State of Matter**<br>How did the states of matter compare between the reactants and products? | | | |

Name______________________________ Date________________

**Comparing Properties of Reactants and Products** *(continued)*

## KEEPING TRACK OF THE MASS

For a very long time, scientists have kept track of the mass of the reactants and products in chemical reactions. Although many scientists kept track of the masses, Antoine Lavoisier is credited with discovering the law of conservation of matter in 1785. The law of conservation of matter is seemingly simple and *very important:* Matter can neither be created nor destroyed. Plainly stated, the mass of the reactants you start with is always equal to the mass of the products you end up with. The mass doesn't change; the atoms are just rearranged in a chemical reaction. We have the exact same number of atoms at the end of the reaction as we had at the beginning.

If that law is true, why wasn't the mass of the reactants equal to the mass of the products in the baking soda and vinegar reaction? Were you able to measure *everything?* What about the gas in the bubbles? If you didn't measure the mass of the gas in the bubbles, some of the mass escaped into the air. To solve the problem, do the reaction in a nonlatex glove.

## GIANT HAND REACTION

This investigation is a repeat of the baking soda and vinegar reaction. This time, the reaction is going to take place in a nonlatex glove so that the gas can be captured. *Keep the fingers of the glove pointed down until after the opening of the glove is tied shut.*

1. Mass a nonlatex glove.
2. Put 5 g of baking soda in one finger of the glove. Twist the finger to seal it off so vinegar won't mix with the baking soda too early. One group member should pinch the glove where it is twisted to keep it from unwinding.
3. Carefully put the vinegar in another finger of the glove. Knot the opening of the glove the same way you would a balloon.
4. Turn the glove upside down so the fingers point up and release the baking soda.
5. After the reaction has finished, mass all of the products while they are still in the glove. Subtract the mass of the glove to find the mass of the products.

Name________________________ Date______________

## Comparing Properties of Reactants and Products (*continued*)

| | Reactants | | | → | Products | | | | |
|---|---|---|---|---|---|---|---|---|---|
| | Baking Soda | + | Vinegar | → | Sodium Acetate | + | Carbon Dioxide | + | Water |
| Mass | | | | | | | | | |
| **Total Mass** | | | | | | | | | |

How does the mass of the reactants compare to the mass of the products?

Were you able to measure the mass of all of the products?

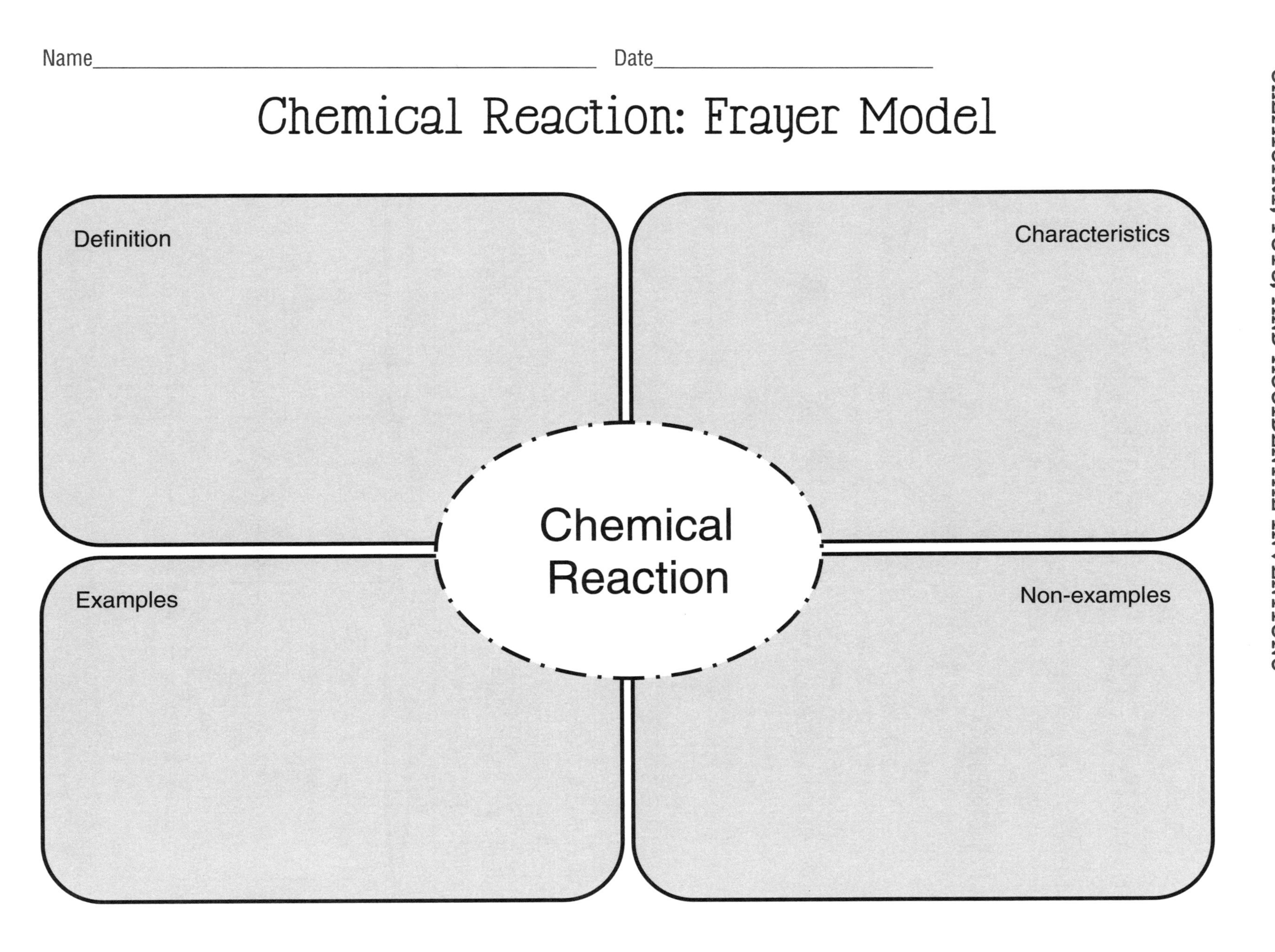
Name
Date
Chemical Reaction: Frayer Model
Definition
Characteristics
Chemical Reaction
Examples
Non-examples

Name____________________ Date____________

# Was There a Chemical Reaction?

Watch this lead iodide video of two solutions being combined *(http://bit.ly/1PVX02v)*. The data for the substances in the video are in the table below.

QR CODE: Lead Iodide Video

## LEAD IODIDE VIDEO DATA

| | Reactants | | | → | Products | | |
|---|---|---|---|---|---|---|---|
| | Sodium Iodide | + | Lead Nitrate | → | Lead Iodide | + | Sodium Nitrate |
| Soluble in Water | Yes | | Yes | | No | | Yes |
| State of Matter | Liquid solution | | Liquid solution | | Solid | | Liquid solution |
| Color | Colorless in solution | | Colorless in solution | | Yellow | | Colorless in solution |
| Mass | 22.63 g | | 25.00 g | | 34.80 g | | 12.83 g |
| **Total Mass** | | | | | | | |

According to the data above about the reactants and products, did a chemical reaction occur? Fill the beaker below with evidence to support your answer.

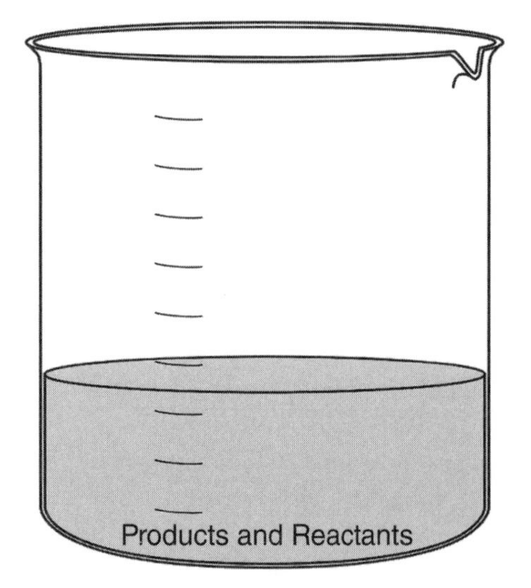


Was the law of conservation of matter observed in the video? Fill the beaker below with evidence to support your answer.

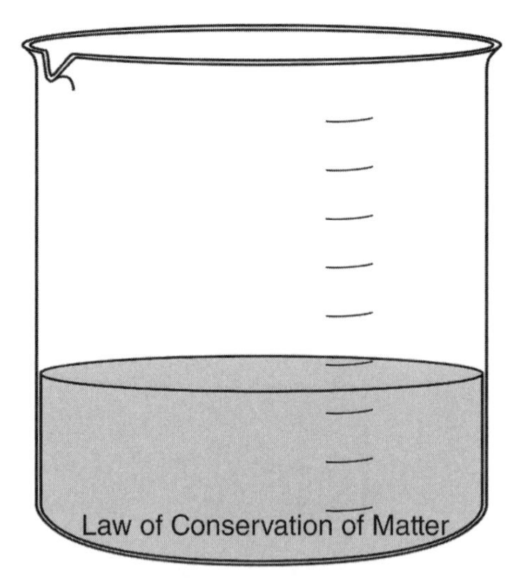

Name______________________________ Date______________________

# Gakish Recipe Cards

## Powder Gakish

### Ingredients

- 5 tbs of baby powder
- 20 ml white glue

### Directions

**Remember to collect data!**

1. Use a tablespoon to carefully measure the baby powder.
2. Use a graduated cylinder to carefully measure the glue.
3. Pour the glue into a beaker or plastic cup.
4. Add the baby powder to the glue and stir.
5. When the product is puttylike, knead it until smooth.

## Laundry Detergent Gakish

### Ingredients

- 35 ml laundry detergent
- 60 ml white glue

### Directions

**Remember to collect data!**

1. Use a graduated cylinder to carefully measure the laundry detergent.
2. Use a graduated cylinder to carefully measure the glue.
3. Pour the glue into a beaker or plastic cup.
4. Add the laundry detergent to the glue and stir.
5. When the product is puttylike, knead it until smooth.

## Chapter 10

# Nature's Light Show: It's Magnetic!

## OVERVIEW

This unit develops an understanding of magnetism—magnetic fields, in general, and Earth's magnetic field, in particular. The unit is situated in the context of understanding the Northern and Southern Lights (aurora borealis and aurora australis, respectively). Students engage in hands-on investigation and read a nonfiction text to answer the question, *What causes Earth's aurora?*

This unit assumes that students know and can explain that magnets attract objects made of certain metals (iron, nickel, or cobalt) but do not attract other types of materials. They should also understand that the poles of two magnets may attract or repel each other, depending on which poles are brought near to each other. If students do not have sufficient experience with magnets to explain these concepts, provide time to explore with magnets and a variety of materials before beginning the unit.

## OBJECTIVES

- Draw and describe Earth's magnetic field.
- Explain the phenomenon of the aurora.
- Summarize a text, including the central ideas and supporting details.
- Synthesize information gained from investigations and text.

## STANDARDS ALIGNMENT

### *Next Generation Science Standards* (NGSS Lead States 2013)

#### PS2.B: TYPES OF INTERACTIONS

- Forces that act at a distance (electric, magnetic, and gravitational) can be explained by fields that extend through space and can be mapped by their effect on a test object (a charged object, or a ball, respectively). (MS-PS2)

### *Common Core State Standards, English Language Arts* (NGAC and CCSSO 2010)

#### GRADES 6–12 LITERACY IN HISTORY/SOCIAL STUDIES, SCIENCE, AND TECHNICAL SUBJECTS

- CCSS.ELA-LITERACY.RST.6-8.2: Determine the central ideas or conclusions of a text; provide an accurate summary of the text distinct from prior knowledge or opinions.

- CCSS.ELA-LITERACY.RST.6-8.3: Follow precisely a multistep procedure when carrying out experiments, taking measurements, or performing technical tasks.
- CCSS.ELA-LITERACY.RST.6-8.7: Integrate quantitative or technical information expressed in words in a text with a version of that information expressed visually (e.g., in a flowchart, diagram, model, graph, or table).
- CCSS.ELA-LITERACY.RST.6-8.9: Compare and contrast the information gained from experiments, simulations, video, or multimedia sources with that gained from reading a text on the same topic.
- CCSS.ELA-LITERACY.WHST.6-8.9: Draw evidence from informational texts to support analysis, reflection, and research.

## TIME FRAME

- Ten 45-minute class periods

## SCIENTIFIC BACKGROUND INFORMATION

Magnetism is a noncontact force, meaning that a magnetic object can exert a force over a second object without touching it. This is due to an area around a magnetic object called a *magnetic field.* A *field,* in science, is defined as a region of space in which a force acts.

Magnetic fields can be visualized by magnetic field lines, which indicate the strength and direction of the magnetic force. Iron filings and compasses can be used to demonstrate the magnetic field of a permanent magnet. It is important to remember that these lines are simply a representation of the actual magnetic field surrounding a permanent magnet.

Earth has a magnetic field, which is caused by the movement of metallic elements in its outer core. This movement generates an electric current, which creates a magnetic field. Earth acts as a giant bar magnet, although the magnetic poles are not in the same location as the geographic North Pole and South Pole. Instead, the magnetic poles are currently located approximately 11 degrees from the geographic poles. Earth's magnetic poles slowly drift, and actually reverse every 300,000 years or so.

Earth's magnetic field allows us to use compasses (which have tiny magnetic needles) to determine magnetic north and magnetic south and orient ourselves on Earth. However, remember that opposite poles of a magnet attract. This means that when a compass's needle points geographically north, it is actually pointing to the magnetic South Pole, and vice versa.

Earth's magnetic field creates a kind of bubble, known as the *magnetosphere,* around the planet. The magnetosphere protects Earth from the *solar wind,* or a stream of electrically charged particles emitted from the Sun. Some of the particles from the solar wind move along Earth's magnetic field lines and are directed toward the North Pole and South Pole. When these charged particles enter Earth's atmosphere, they crash into gas molecules, releasing energy in the process. This energy creates the colorful displays of light seen in the polar regions, known as the *aurora borealis* in the Northern Hemisphere and the *aurora australis* in the Southern Hemisphere.

## MISCONCEPTIONS

A wide variety of misconceptions have been documented about magnetism and Earth's magnetic field. For the purposes of this unit, we share some misconceptions in Table 10.1.

### TABLE 10.1. COMMON MISCONCEPTIONS ABOUT MAGNETISM AND EARTH'S MAGNETIC FIELD

| Common Misconception | Scientifically Accurate Concept |
|---|---|
| A magnetic field is a pattern of lines surrounding a magnet. | A magnetic field is an area in which the force of magnetism acts. Field lines are a way to visualize a magnetic field. |
| The magnetic and geographic poles of Earth are located in the same place. | The magnetic poles are located approximately 11 degrees from the geographic poles. The location of the magnetic poles changes slowly. Magnetic north and magnetic south have reversed throughout Earth's history. |
| The magnetic pole of Earth in the Northern Hemisphere is a north pole, and the pole in the Southern Hemisphere is a south pole. | The geographic North Pole is a magnetic south pole, and the geographic South Pole is a magnetic north pole. |

*Source:* American Institute of Physics 2016; Hapkiewicz 1992.

## MATERIALS

### For the Small-Group Work

- Plastic shoebox
- Index card (3 × 5 in.)
- Iron filings
- Bar magnet
- Cardboard cut to the size and shape of the bar magnet
- 16 compasses
- Printer paper
- 16 oz. plastic bottle (such as a water or soda bottle)
- Manila folder
- Piece of tissue paper
- Cow magnet (available from science supply stores and online retailers)
- Printed infographic (see "Nonfiction Texts" section)
- 10 cm (4 in.) piece of plastic straw
- 2 steel straight pins
- Magnet

### For the Work in Pairs

- Bar magnet
- Compass

### For Each Student

- Computer, laptop, or tablet with Adobe Flash Player and internet access

### General Supplies

- Tape (transparent, masking, or duct)
- Sewing thread
- Paper grocery bags made from recycled materials (one per planet; see Expand phase)
- Plastic grocery bags
- Dead, D-size battery (one per planet; see Expand phase)
- 2 ceramic donut magnets (1 and 1/8 in. diameter)
- 3 rubber bands (size 64 preferred)
- Marker
- Goggles (sanitized, indirectly vented chemical splash)
- Nonlatex gloves
- Aprons

### Optional Material

- Document camera

## SUPPORTING DOCUMENTS

- "Investigating Magnets I" lab sheet
- "Investigating Magnets II" lab sheet
- "What Is a Magnetic Field" nonfiction text
- "Investigating Magnets III" lab sheet
- "Investigating Earth's Magnetic Field" lab sheet
- "Modeling Earth's Magnetic Field" lab sheet
- "Summarizing 'Earth's Magnetic Field'" graphic organizer
- "The Aurora: Fire in the Sky" nonfiction text
- "Pulling It All Together" graphic organizer
- "Planning My Infographic" graphic organizer
- "Planet Data Collection" data sheet
- "Planet Information Cards"

## NONFICTION TEXTS

- "The Aurora: Fire in the Sky" by Stephen Whitt, Beyond Penguins and Polar Bears, 2008, *http://beyondpenguins.ehe.osu.edu/issue/polar-patterns-day-night-and-seasons/the-aurora-fire-in-the-sky;* Flesch-Kincaid reading level 5.2.

This nonfiction article has two versions that are written at lower reading levels, in addition to the version found at the end of this unit.

QR CODE 1

- "Earth's Magnetic Field" by Randy Russell, Windows to the Universe, 2010, *http://bit.ly/1Q1JCpu;* Flesch-Kincaid reading level 6.4. (QR code 1)

This article describes Earth's magnetic field and how it is created. The information listed above is for the intermediate version of the article. The article is also available in beginner and advanced levels, and all levels are available in Spanish.

- Various infographics by Kids Discover, *http://bit.ly/1oeoMNK.* (QR code 2)

QR CODE 2

Many examples of infographics are available for free download from the Kids Discover website. Students will need to examine an infographic during the Explain phase.

The following nonfiction passages are provided at the end of the unit:

- "The Aurora: Fire in the Sky" by Stephen Whitt (Flesch-Kincaid reading level 5.2)
- "What Is a Magnetic Field?" (Flesch-Kincaid reading level 8.2)

## SAFETY CONSIDERATIONS

- Have students wear goggles, gloves, and aprons during all phases of the investigation, including during setup and cleanup.
- Review safety information in safety data sheets for iron filings with students (see Appendix 4).
- Remind students to use caution in working with iron filings; they are sometimes sharp and can puncture skin.
- Remind students to be careful and not breathe in the dust from the iron filings.
- Make sure students have appropriate procedures for cleanup and disposal of iron filings.
- Have students wash hands with soap and water after completing the investigation.

## LEARNING-CYCLE INQUIRY

### Engage

QR CODE 3

In this phase, students will view and discuss videos of the Northern and Southern Lights.

1. Show students the video "Aurora Borealis, February 18 2014, Fairbanks, Alaska" *(http://safeshare.tv/v/6Dz1U9P0i-I;* QR code 3*)*. Ask students to share what they observed in the video. Explain that this phenomenon is called the *aurora borealis,* or Northern Lights.

QR CODE 4

2. Show students the video "Aurora Australis—Southern Lights, 15 July, 2012" *(http://safeshare.tv/v/okPhl45WFMk;* QR code 4*)*. Explain that the same phenomenon, when it occurs in the Southern Hemisphere, is called the *aurora australis,* or Southern Lights.
3. Show students the video "Aurora from ISS Orbit" *(http://go.nasa.gov/1wnVF6g;* QR code 5*)*. Explain that this video was taken from the International Space Station. Ask students to share what they observed in the video.

QR CODE 5

4. Finally, show students the video "NASA IMAGE—Spacecraft Pictures Aurora" *(http://safeshare.tv/v/8_V7c756Oro;* QR code 6*)*. Explain that these images were taken from a satellite. Ask students to share what they observed in the video.
5. Explain that the Northern and Southern Lights, or aurora, are most often observed at high latitudes. Share that to native peoples living in the Arctic region, the aurora was often a focus of wonder and fear. Explain that scientists, too, wondered about this seemingly mysterious phenomenon. Introduce the unit question, *What causes Earth's aurora?*

QR CODE 6

***Assess this phase:*** Student participation in class discussions serves as a formative assessment. No other assessment is needed at this time.

### Explore

In Part I of this phase, students engage in direct explorations of the magnetic fields of bar magnets. They also read nonfiction texts to build their understanding of a dipole magnetic field. In Part II, students engage in direct explorations of Earth's magnetic field using bar magnets and an indirect exploration of Earth's magnetic field using a web-based simulation. By comparing sketches from the two explorations, they determine that Earth behaves much like a bar magnet.

#### PART I: UNDERSTANDING MAGNETIC FIELDS

1. Ask students to share what they know about magnets. If student responses indicate a lack of experience with magnets, distribute magnets and a variety of magnetic and nonmagnetic materials and allow them to explore. It will be helpful to provide

students with items made of magnetic and nonmagnetic metals, as well as other nonmagnetic items.

2. Distribute copies of the "Investigating Magnets I" handout to students. Review the directions with students before they begin the investigation. Instruct students to be careful to keep the iron filings away from the magnets, because they are difficult to remove. (*Note:* Students may have difficulty creating the field lines with the iron filings. Circulate around the room and assist students in gently tapping the sides of the shoeboxes so that the field lines are visible.)
3. Engage students in a discussion of their observations. Use guiding questions such as the following to focus student attention on the magnetic field:
    - What did you observe when you placed the iron filings on top of the cardboard?
    - What did you observe when you placed the iron filings on top of the bar magnet?
    - Why do you think the iron filings behaved the way that they did?
    - Why do you think there was a difference in the behavior of iron filings between the cardboard and bar magnet?

    (*Note:* If you have a document camera, project student sketches during this class discussion.)
4. Pass out the compasses and allow students to examine them. Ask students to share what they know about compasses. Explain to students that a compass is made of a tiny magnetic needle, and that they will explore how the compass changes when it interacts with a bar magnet.
5. Distribute the "Investigating Magnets II" handout. Review the directions with students before they begin the investigation. (*Note:* This investigation is best completed in pairs.)
6. Engage students in a discussion of their observations. Use guiding questions such as the following to focus student attention on the magnetic field lines:
    - How do the lines compare to the ones that the iron filings made?
    - What do you notice about the direction in which the lines point?
    - What do you think these lines represent?
    - Why do you think the compass behaved the way it did around the magnet?
7. Projecting one pair of students' work or a comparable picture during this discussion will be a helpful reference.
8. Have students read and discuss "What Is a Magnetic Field?" Ask them to share what connections they can make between their investigations and the text.

9. Discuss and define *magnetic field*. A suggested definition is the area around a magnet in which the force of a magnet can act, but allow students to define the term in their own words. Post this definition in a visible place for the duration of the unit.

### PART II: UNDERSTANDING EARTH'S MAGNETIC FIELD

1. Explain to students that they will visualize the magnetic field of a bar magnet in a different way—with compasses.
2. Distribute "Investigating Magnets III" to students. Review the directions with students before they begin the investigation. (*Note:* This investigation can be completed in pairs or small groups, depending on how many compasses are available. It could also be set up as a demonstration if needed.)
3. Engage students in a discussion of their observations. Use guiding questions such as the suggestions below to focus student attention on the magnetic field:

    - What do you notice about the direction in which the compass needles point?
    - Why do you think the needles point the way they do?
    - Can you make a connection between the magnetic field you drew and this sketch?

    If you have a document camera, project student sketches during this class discussion. Use the discussion questions to guide students to an understanding that the compass needles reflect the force of the magnetic field around the magnet.

QR CODE 7

4. Display and preview the interactive simulation "Earth's Magnetic Field Interactive" (*http://bit.ly/1jv3iVM;* QR code 7). Explain to students that they will use this simulation to learn about Earth's magnetic field. Distribute copies of the handout, "Investigating Earth's Magnetic Field," and review the directions before allowing students to begin to work. (*Note:* This simulation is best suited for individual work, although you can certainly use other instructional pairings as needed.)
5. Engage students in a discussion of their observations. Use guiding questions such as the following to focus student attention on Earth's magnetic field:

    - What do you notice about the direction in which the compass needles point?
    - Why do you think the needles point the way they do?
    - Can you make a connection between the magnetic fields you drew in other investigations and this sketch?
    - How does Earth's magnetic field compare to that of a bar magnet?

6. It may be helpful to display and manipulate the compass in the simulation during the discussion. Use the discussion questions to guide students to an understanding

that the compass needles reflect the force of the magnetic field around Earth and that Earth's magnetic field resembles that of the bar magnets they previously investigated.

7. Next, students will create a 3-D model of a magnetic field to help them better visualize Earth's magnetic field. Distribute the "Modeling Earth's Magnetic Field" handout and review directions with students before they begin the activity. (*Note:* This activity is best conducted in pairs or small groups, although it can be also done as a whole-class demonstration.)
8. Engage students in a discussion about their observations. Use guiding questions such as the following to help build student understanding of Earth's magnetic field:
   - What did you observe when you added the magnet to the bottle of iron filings?
   - How does what you see compare to your observations of magnetic fields in our previous investigations?
9. Direct students to read "Earth's Magnetic Field." Because this text is available at three reading levels, you may choose to break students into small groups to read and discuss.
10. Explain to students that they will learn how to write a summary paragraph of the article. Distribute the handout "Summarizing 'Earth's Magnetic Field.'" Review the directions with students, and model how to complete the first row of the organizer with the main idea of the first paragraph (Earth has a magnetic field) and key (supporting) details. (*Note:* If students have not had sufficient practice with identifying the main idea and supporting details or with summarizing, we recommend you continue to work through the handout with students.) You may also release students who are comfortable working independently, and continue teacher-directed instruction with a small group.
11. Draw student attention to the word *magnetosphere.* Ask students to think-pair-share about a definition for this word, using what they have observed in the investigations and learned in the reading. After hearing student responses, settle on a class definition and post it in a prominent place for the duration of the lesson.
12. Remind students of the initial question, *What causes Earth's aurora?* Distribute copies of the nonfiction text "The Aurora: Fire in the Sky" to students, and explain that their purpose in reading is to determine the answer to this question.
13. Discuss the reading with students, using guiding questions to probe their comprehension of the article. Some examples of questions for discussion include the following:
    - What is the solar wind?
    - Why is the solar wind important?

QR CODE 8

QR CODE 9

- How does electricity relate to the solar wind?
- How does the solar wind interact with Earth's magnetic field?
- What causes the aurora?

Students may need additional time to develop an understanding of how the interaction between the electrically charged solar wind and Earth's magnetic field produces the aurora. We recommend showing students the video "Solar Wind's Effect on Earth" (*http://bit.ly/1Sp4m0A;* QR code 8) and the animation "Earth's Magnetic Field to Aurora" (*http://go.nasa.gov/211V0JC;* QR code 9).

***Assess this phase:*** Students' completion of handouts for each investigation and participation in class discussions are key sources of formative assessment. By the end of Part I, students should be increasingly familiar with the concept of a magnetic field. By the end of Part II, students should be able to explain what Earth's magnetic field is and identify the solar wind and the magnetic field as the contributing causes of the aurora. If they are unable to do so, they will not be successful with the instructional activities in the Explain phase. If students are not ready to proceed, return to the activities in a teacher-directed manner and consider supplementing the lesson with video clips and other resources to build the requisite knowledge.

## Explain

In this phase, students will complete a graphic organizer to synthesize the knowledge they have gained in the previous phases. They will use a free online tool to create an infographic to answer the question, *What causes Earth's aurora?*

1. Distribute copies of the "Pulling It All Together" graphic organizer to students. Explain that they will use this organizer to record what they have learned about Earth's magnetic field and the solar wind, and then they will synthesize (combine) that information to explain what causes the aurora. Students should have access to all papers and materials from investigations as they complete this organizer. Students should submit this organizer for your review before proceeding to step 2.
2. Explain to students that they will be creating infographics to answer the unit question, *What causes Earth's aurora?* Introduce students to infographics by giving small groups of students a printed example of an infographic and asking them to discuss it. Use guiding questions such as the following to help students dig deep into the meaning behind the infographic:
   - What information is this infographic trying to communicate?
   - What data are represented in the infographic?
   - How did the creator of this infographic choose to represent the data?

- How do the images and text work together to communicate information?

3. Ask small groups to share their infographics and responses to the questions. Continue discussing each example as a whole class, focusing on how data and information are represented visually and in text.
4. Tell students that they will now create their own infographic to explain what causes the aurora. Students should use the results from their observations and reading to create this infographic. Distribute the "Planning My Infographic" graphic organizer to students to help them get started.

   Many options exist for creating infographics online. We recommend using Google Drawings, which is part of the Google Apps suite, because it is a free and easy-to-use application. Students can upload photos of their own work from previous investigations or search the web for Creative Commons–licensed images. They can edit and layer these images to create their own infographic. If you use Google Classroom, you can easily create folders in which the entire class can save its work.
5. When students have completed their infographics, allow time for them to share their work with the class. This may be accomplished by projecting infographics for the whole class to view, or by conducting a gallery walk. (*Note:* For an explanation of the gallery walk strategy, see page 342 in Appendix 2.)

***Assess this phase:*** Completed "Pulling It All Together" graphic organizers serve as the final formative assessment for this unit. It is essential that you review these organizers before allowing students to create their infographics. In the organizer, students should demonstrate an understanding of magnetic fields, Earth's magnetic field, and, to a lesser extent, the solar wind. Students should be able to explain that the solar wind is made of electrically charged particles that are released by the Sun and travel through space to Earth. Students should also be able to explain that as these charged particles travel toward the North Pole and South Pole along Earth's magnetic field lines, they interact with gas molecules and release the energy that creates the aurora. If student work is incomplete or incorrect, have students return to the nonfiction text, videos, and animations for further reading and review.

Student-created infographics serve as summative assessment for this unit. The science content expressed should be the same as in the "Pulling It All Together" graphic organizer. On the infographics, look for students' abilities to convey this understanding through both their images and text.

## Expand

In this phase, students will test "planets" for the presence or absence of a magnetic field and will use that information to predict whether that planet has auroras.

***Advance preparation:*** Complete the steps that follow before beginning the Expand phase.

1. Build four "planets' of varying sizes from paper bags stuffed with plastic bags and held together with rubber bands, following the directions given at *http://bit.ly/1oeXvdQ*. Of the four planets, at least one should be prepared exactly according to these directions with a magnetic core. The others should be prepared with the dead battery in the center, but no magnets—meaning that they lack a magnetic core. Label these planets A–D, noting which have a magnetic core and which do not. (*Note:* Depending on your class size, you may wish to construct several identical copies of each planet so that students are not waiting to begin gathering data on each planet.)
2. Build enough magnetometers for each small group to have access to one, according to the directions at *http://bit.ly/1QzFeOI.*
3. Copy the "Planet Information Cards" to accompany each model planet.

After preparations are made, begin the Expand phase. The steps are as follows:

1. Explain to students that other planets besides Earth experience auroras. Explain to students that they will collect data about four newly discovered planets (Planet A, Planet B, Planet C, and Planet D) and that they will use that data to predict whether each planet has an aurora. Distribute copies of the "Planet Data Collection" handout to students. Review the directions with them and demonstrate how to use the magnetometer with a bar magnet.
2. Circulate around the room while students work, using the following guiding questions to gauge student understanding:
    - What are you learning about this planet?
    - How is this planet like the others? How is it different?
    - How will you determine which planets have an aurora?
3. Engage students in a whole-class discussion about their findings. Discuss which planets can be expected to have auroras, and discuss how students arrived at this decision.

***Assess this phase:*** Completed data collection sheets and student participation in the whole-class discussion serve as the final summative assessments for this unit. Students should be able to correctly predict which planets have an aurora and explain that the presence of the magnetic field is needed for this to happen.

## REFERENCES

American Institute of Physics. 2016. Children's misconceptions about science. *http://amasci.com/miscon/opphys.html.*

Hapkiewicz, A. 1992. Finding a list of science misconceptions. *MSTA Newsletter* 38 (Winter): 11–14.

National Governors Association Center for Best Practices and Council of Chief State School Officers (NGAC and CCSSO). 2010. *Common core state standards.* Washington, DC: NGAC and CCSSO.

NGSS Lead States. 2013. *Next Generation Science Standards: For states, by states.* Washington, DC: National Academies Press. *www.nextgenscience.org/next-generation-science-standards.*

Name________________________________________ Date______________________

# Investigating Magnets I

## MATERIALS

- 1 plastic shoebox
- 2 pieces of white printer paper
- Iron filings
- Bar magnet
- Rectangular piece of cardboard
- Goggles
- Nonlatex gloves
- Aprons

## SAFETY REMINDERS

- Wear goggles, gloves, and aprons during all phases of the investigation, including during setup and cleanup.
- Be careful when working with iron filings; they can be sharp and cut your skin.
- Be careful not be breathe in the dust from these filings.
- Follow your teacher's directions for cleanup and disposing of the iron filings.
- Wash hands with soap and water after completing the investigation.

## PROCEDURE

1. Trace the piece of cardboard in the center of one piece of printer paper.
2. Place the paper inside the shoebox so it lays flat on the bottom of the box.
3. Place the piece of cardboard on the table. Carefully place the shoebox on top of the cardboard. Try to line up the cardboard with the shape you traced.
4. Add 1 tbsp of iron filings to the box. Gently tap or shake the box so that the filings spread out across the paper.
5. What do you notice about the iron filings? Sketch your observations in the space below.

Name______________________________ Date__________________

## Investigating Magnets I (*continued*)

6. Carefully pour the iron filings out of the box. Remove the paper.
7. Trace the bar magnet in the center of one piece of printer paper. Be sure to label the north and south poles of the magnet in your drawing.
8. Place the printer paper inside the shoebox so it lays flat on the bottom of the box. Carefully place the shoebox on top of the magnet. Try to line up the magnet with the shape you traced on the paper.
9. Add 1 tbsp of iron filings to the box. Gently tap or shake the box so that the filings spread out across the paper.
10. What do you notice about the iron filings? Sketch your observations in the space below.

What differences did you observe in the appearance of the iron filings with the cardboard and the bar magnet?

______________________________

______________________________

______________________________

______________________________

Why do you think this difference occurred?

______________________________

______________________________

______________________________

______________________________

Name________________________________________ Date____________________

# Investigating Magnets II

## MATERIALS

- 2 pieces of white printer paper
- Tape
- Bar magnet
- Compass

In this investigation, you will move a compass around a bar magnet and use the compass to draw lines around the magnet. Follow these instructions:

1. Tape the two pieces of paper together.
2. Place the bar magnet in the middle of the paper. Trace it, and label the north and south poles above and below the magnet.
3. Draw a dot somewhere near the magnet and put the compass over the dot. Make sure the compass is centered on top of the dot.
4. Draw a dot at the location of the arrowhead of the compass needle.
5. Move the compass to sit on the new dot. Again, make sure the compass is centered on top of the dot.
6. Draw a dot at the location of the arrowhead of the compass needle.
7. Take the compass off the paper. Draw lines connecting the dots. Add arrows showing the direction in which the compass points.
8. Continue with steps 3–7 until your line touches the magnet or the edge of the paper.
9. Choose another spot near the magnet and repeat steps 3–8.
10. Repeat until you have lines surrounding the magnet.

Compare your drawing to your sketch in "Investigating Magnets I." What do you notice?

______________________________________________________________________

______________________________________________________________________

______________________________________________________________________

______________________________________________________________________

______________________________________________________________________

______________________________________________________________________

Name________________________________________ Date______________________

# What Is a Magnetic Field?

Magnetism is an invisible force caused by the unique properties of certain materials. You know that all matter is made up of atoms. In the atoms of most materials, electrons spin in different, random directions. In magnetic materials, however, the atoms are arranged so that their electrons all spin in the same direction. The spin of these electrons creates a magnet with two poles, a north-seeking pole and a south-seeking pole.

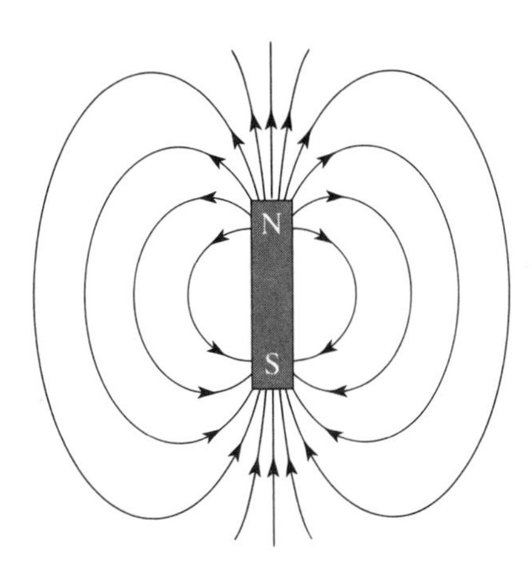


If you take a paperclip and move it close to a magnet, there will be a point where the paperclip is attracted to the magnet. That point is the edge of the magnet's **magnetic field.** A magnetic field is an area around a magnet in which its force can act on an object. A magnet's magnetic field is caused by the magnetic force flowing from the North Pole to the South Pole.

Although a magnetic field is invisible, we can use magnetic materials to help visualize it. We can use small bits of metal called *iron filings* to show a magnetic field. If you place a bar magnet under a flat surface and sprinkle iron filings on the surface, the filings will arrange themselves along magnetic field lines. We can also use a compass to help use "see" the magnetic field.

Name______________________________ Date______________

# Investigating Magnets III

## MATERIALS

- Bar magnet
- 16 compasses

## PROCEDURE

1. Place the bar magnet on the table. Arrange the compasses around the magnet as shown in the diagram below.
2. Observe the compasses and the direction in which each needle is pointing. Draw the needle in each compass in the diagram below to record your observations.

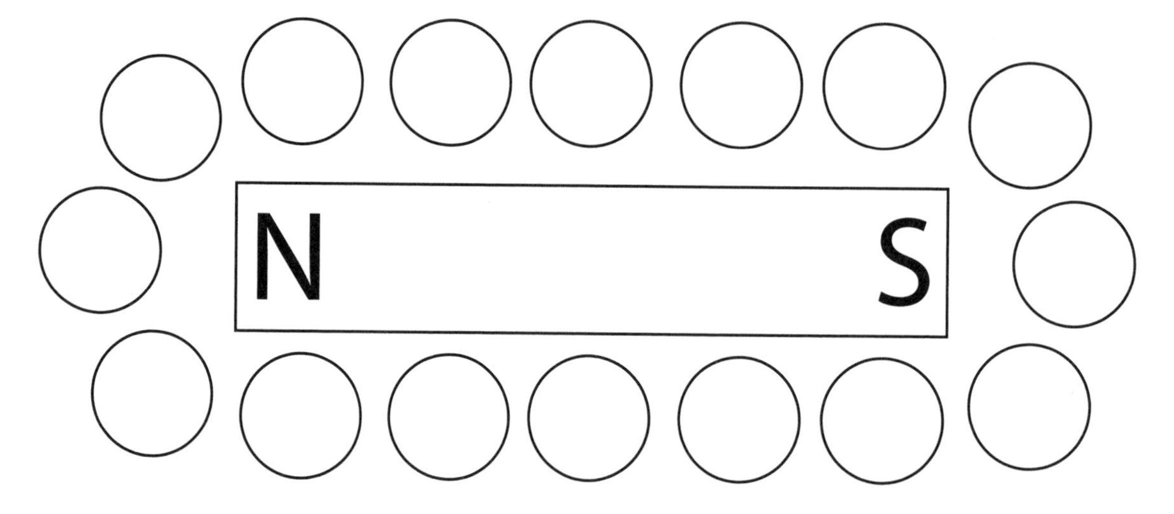


What do you notice about the direction in which the compass needles are pointing?

______________________________

______________________________

______________________________

______________________________

______________________________

______________________________

Name________________________________________ Date______________________

# Investigating Earth's Magnetic Field

To complete this investigation, you will need only a computer.

## PROCEDURE

QR CODE: "Earth's Magnetic Field Interactive"

1. Open the "Earth's Magnetic Field Interactive" at *http://bit.ly/1jv3iVM.*
2. Drag the compass around the picture of Earth to see how the position of the needle changes.
3. Draw the needle in each compass in the diagram below to record your observations.

What do you notice about the direction in which the compass needles are pointing?

______________________________________________________________________

______________________________________________________________________

______________________________________________________________________

______________________________________________________________________

______________________________________________________________________

______________________________________________________________________

Name______________________________ Date__________________

# Modeling Earth's Magnetic Field

## MATERIALS

- Cow magnet
- 16 oz. plastic bottle
- 2 tbsp of iron filings
- Manila folder
- Tape
- Piece of tissue paper
- Goggles
- Nonlatex gloves
- Aprons

## SAFETY REMINDERS

- Wear goggles, gloves, and aprons during all phases of the investigation, including during setup and cleanup.
- Be careful when working with iron filings; they can be sharp and cut your skin.
- Be careful not be breathe in the dust from these filings.
- Follow your teacher's directions for cleanup and disposing of the iron filings.
- Wash hands with soap and water after completing the investigation.

## PROCEDURE

1. Cut the manila folder in half, and roll up one half tightly into a tube that is about the diameter of the cow magnet. The tube should be slightly longer than the bottle. Tape the tube so it stays rolled up.
2. Seal off one end of the tube by taping it.
3. Crumple the tissue paper and stuff it down into the tube by pushing it down with a pencil.
4. Carefully pour the iron filings into the bottle. Pour enough filings to coat the bottom with a layer ¼ in. thick.
5. Insert the tube into the bottle. Use paper and tape to seal the bottle opening around the tube.
6. Slide the cow magnet into the tube. It should rest on the crumpled tissue paper and not reach the bottom of the tube.
7. Hold the magnet in place with the pencil and carefully shake the bottle. Observe what happens to the iron filings. Sketch your observations.

Name________________________________ Date____________________

# Summarizing "Earth's Magnetic Field"

Complete this organizer after reading the article "Earth's Magnetic Field." Then, use the information in the organizer to write a summary of the article.

| Main Idea | Key Details |
|---|---|
| | 1.<br>2. |
| | 1.<br>2. |
| | 1.<br>2. |
| | 1.<br>2. |

Name______________________ Date______________

Summarizing "Earth's Magnetic Field" (*continued*)

| Main Idea | Key Details |
|---|---|
| | 1.<br>2. |
| | 1.<br>2. |
| Summary of article | |

Name________________________ Date________________

# The Aurora: Fire in the Sky

*by Stephen Whitt*

The Northern Lights (or the Southern Lights, if you're in the Southern Hemisphere) are eerie, multicolored streaks and shapes that appear in the night sky, as if from nowhere. To find out where they come from, we'll have to take a little trip. Are you ready?

Imagine you are on the Sun. The Sun's temperature is much too hot for anything alive. But you aren't alive. You are a tiny **particle** so small that you can't be seen in even the most powerful microscope.

Now, imagine that you are hurled away from the Sun. Believe it or not, this actually happens all the time. The Sun sends out streams of tiny particles every second. We call this stream the **solar wind**. Heat causes the solar wind. The Sun is so hot that particles fly off its surface, a little like steam rising from a hot bowl of soup.

Imagine you're a part of this solar wind. You're flying away from the Sun faster than the fastest spaceship. Directly ahead of you is Earth, a pretty blue-white ball. You're moving fast, but Earth is still far away. It takes you a little over four days to make the trip to Earth.

What happens when you reach Earth? To find out, let's leave the solar wind for a moment and travel back in time to meet a scientist and explorer named Kristian Birkeland.

Kristian Birkeland wanted to understand the aurora (another name for the Northern and Southern Lights). The mysterious light was often seen near the North and South Poles. It wasn't usually seen closer to the equator. People described the light as a "fire in the sky." But what could it be?

Birkeland had an idea. He knew that Earth was a giant magnet. Like all magnets, Earth has a north magnetic pole and a south magnetic pole. Birkeland led an **expedition** to Norway to measure Earth's **magnetic field.** He found that near the North Pole, the magnetic field lines don't run along Earth's surface the way they do near the equator. Instead, the field lines go almost straight up and down. What could that mean?

Name________________________________________ Date______________________

## The Aurora: Fire in the Sky *(continued)*

Think of Earth as a magnet. Near the middle of the magnet (where Earth's equator would be), the lines of force run right alongside the magnet. But near the North and South Poles, the lines run almost straight into the ends of the magnet.

Kristian Birkeland now knew more about Earth's magnetic field. But he still didn't know what caused the aurora. How were the two things related?

To understand how these things are related, you need to know a little about electricity. Have you ever rubbed your feet across the carpet and then touched something made of metal? If you have, you've felt a shock! You build up an **electric charge** when you rub your feet on the carpet. The charge moves from your finger to the metal when you touch it. This movement is what causes the shock. The particles from the Sun also carry an electric charge. But how does this charge create the aurora?

Here's the key idea. *Electricity and magnets affect each other.* Watch a **compass** during a thunderstorm. You'll see the magnet inside the compass (what we call the needle) moves every time lightning flashes across the sky.

Now, we know that electricity affects magnets. But do magnets affect electricity? Yes! Birkeland showed how by building a magnetic model of Earth. He found that the charged particles traveled along the magnetic field lines. They moved away from the equator and followed the lines to the North and South Poles.

Now, let's go back to those real charged particles flying off the Sun. Just like in the model, the charged particles are pushed by Earth's magnetic field toward the poles. Once they get there, they follow the magnetic field lines down toward the ground. Before the charged particles can get to the ground, though, they smash into **molecule**s in the air. The collisions make the molecules glow with beautiful, bright colors—green, pink, and red. This is the aurora, which is light created by tiny particles from the Sun smashing into Earth's atmosphere at the end of a four-day journey through space. That's quite a trip!

Name______________________________ Date____________________

## The Aurora: Fire in the Sky (*continued*)

## GLOSSARY

**Aurora**—another name for the Northern or Southern Lights

**Compass**—a tool that measures Earth's magnetic field and is used to find directions

**Electric charge**—a measure of the extra positive or negative particles that an object has

**Expedition**—a trip made by a group of people for a particular purpose

**Magnetic field**—the space all around a magnet where the force of the magnet can act

**Molecules**—a grouping of two or more atoms joined together

**Particles**—tiny pieces of matter that make up solids, liquids, and gases

**Solar wind**—electrically charged particles that come from the Sun

---

Name________________________________________ Date______________________

# Pulling It All Together

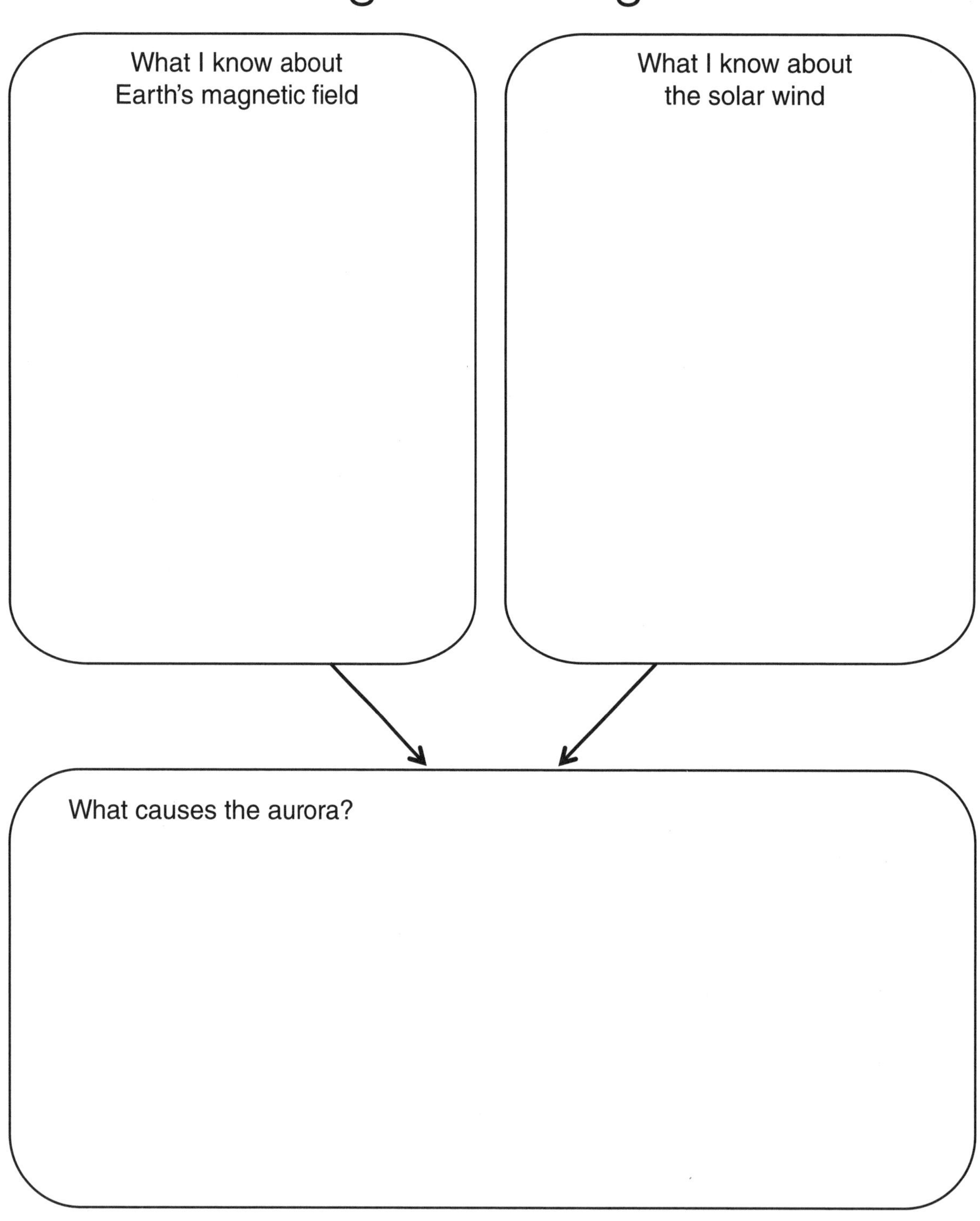

Name_______________________________ Date_______________

# Planning My Infographic

My infographic should answer the question, *What causes Earth's aurora?*

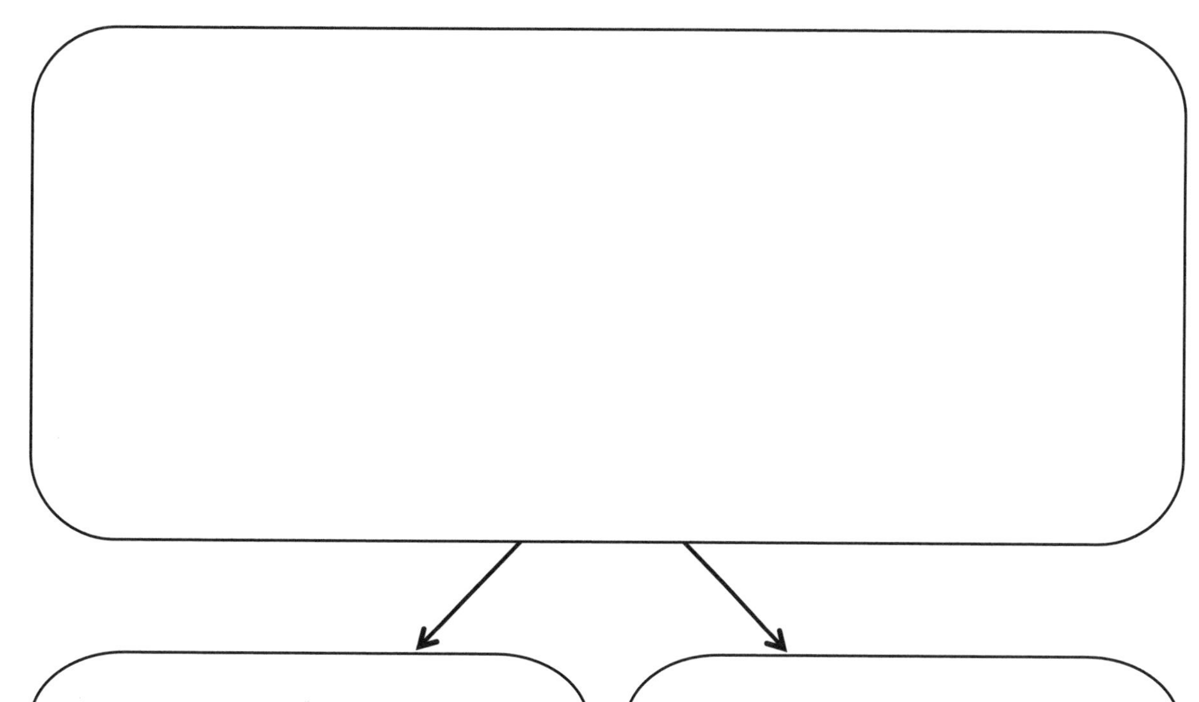

Pictures I want to use

1. ____________________

2. ____________________

3. ____________________

4. ____________________

5. ____________________

What I want to say in writing

Name____________________ Date________________

# Planet Data Collection

Please record the following information about each planet.

| Planet | Distance From the Sun | Features | Magnetic Field (Yes/No) |
|---|---|---|---|
| A | | | |
| B | | | |
| C | | | |
| D | | | |

Use the data to answer the following questions:

1. Which planet(s) would you expect to have auroras?

2. Why did you make this prediction? Explain your thinking.

Name______________________________ Date__________________

# Planet Information Cards

**Planet A**

- Atmosphere
- 4,000 million miles from the Sun
- No known moons

**Planet B**

- Atmosphere
- 5,000 million miles from the Sun
- No known moons

**Planet C**

- Atmosphere
- 6,000 million miles from the Sun
- No known moons

**Planet D**

- Atmosphere
- 7,000 million miles from the Sun
- No known moons

## Chapter 11

# Thermal Energy: An Ice Cube's Kryptonite!

## OVERVIEW

In this unit, students work in small groups to explore conduction and the thermal properties of common materials. They investigate how various materials affect the rate at which ice cubes melt, compare the difference between measuring temperature by touch and with a thermometer, explore and define thermal equilibrium and conduction, draw a relationship between mass and thermal energy, and design a reusable school lunch bag made out of readily available and inexpensive materials. The activities alternate between guided and open inquiry. Throughout the unit, students work with print and online media.

This unit assumes that students know and can demonstrate that matter is made up of atoms, that materials can be described by their properties, that different materials have different properties, and that kinetic energy is associated with the motion of objects, including atoms. Students should also be able to use basic lab equipment, including thermometers, and know how to construct graphs from data. If students lack basic lab skills, plan ahead to insert just-in-time mini-lessons on how to use equipment and how to construct graphs.

## OBJECTIVES

- Recognize that different materials have different thermal properties.
- Distinguish between qualitative and quantitative measurements of temperature.
- Illustrate that thermal energy moves from areas of high heat to areas of low heat.
- Apply knowledge of thermal energy transfer to a design challenge.
- Communicate scientific knowledge in a variety of ways.

## STANDARDS ALIGNMENT

### *Next Generation Science Standards* (NGSS Lead States 2013)

#### PS3.A: DEFINITIONS OF ENERGY

- Temperature is a measure of the average kinetic energy of particles of matter. The relationship between the temperature and the total energy of a system depends on the types, states, and amounts of matter present. (MS-PS3-3 and MS-PS3-4)

#### PS3.B: CONSERVATION OF ENERGY AND ENERGY TRANSFER

- The amount of energy transfer needed to change the temperature of a matter sample by a given amount depends on the nature of the matter, the size of the sample, and the environment. (MS-PS3-4)
- Energy is spontaneously transferred out of hotter regions or objects and into colder ones. (MS-PS3-3)

### Common Core State Standards, English Language Arts (NGAC and CCSSO 2010)

#### GRADES 6–12 LITERACY IN HISTORY/SOCIAL STUDIES, SCIENCE, AND TECHNICAL SUBJECTS

- CCSS.ELA-LITERACY.RST.6-8.3: Follow precisely a multistep procedure when carrying out experiments, taking measurements, or performing technical tasks.
- CCSS.ELA-LITERACY.RST.6-8.4: Determine the meaning of symbols, key terms, and other domain-specific words and phrases as they are used in a specific scientific or technical context relevant to grades 6–8 texts and topics.
- CCSS.ELA-LITERACY.RST.6-8.7: Integrate quantitative or technical information expressed in words in a text with a version of that information expressed visually (e.g., in a flowchart, diagram, model, graph, or table).
- CCSS.ELA-LITERACY.RST.6-8.9: Compare and contrast the information gained from experiments, simulations, video, or multimedia sources with that gained from reading a text on the same topic.

## TIME FRAME

- Six to seven 45-minute class periods

## SCIENTIFIC BACKGROUND INFORMATION

We often think of temperature as how hot or cold something is, and that's not entirely incorrect. It's our common daily understanding of temperature. In daily language, the term *temperature* is a descriptive term used for comparison; it is not exact. The term is biased by our opinions of what we think of as being hot and cold. For instance, we might describe the soup as being too hot. By that, we likely mean that it is unsafe to eat the soup at its current temperature. We are comparing safe and unsafe temperatures for soup. Whereas one person might like his or her soup very hot, another might like it lukewarm. Even when we hear a measured temperature, such as when the weather person tells us the low temperature for the next day is going to be 10°F, we process that information by associating it with cold weather in general—really cold weather! We think of temperature in the context of how it feels.

In science, the term *temperature* is the measure of the average kinetic energy of a substance. It is a more exact use of the concept of temperature. We still compare temperatures, but we don't compare them on the basis of how they feel. We compare temperatures as they are measured with a thermometer. The thermometer removes the bias that exists when we think about temperature in the context of how it feels (The Physics Classroom 2016).

When two objects of different temperatures come into contact, they eventually reach thermal equilibrium. That is, they reach the same temperature. The warmer object cools down, and the cooler object warms up. Energy is transferred from one object to the other. The energy that is transferred is referred to as *heat*. Heat is the flow of energy from one object to another—from the warm object to the cool object. The heat flows until thermal equilibrium has been reached (Nave 2016).

We often talk about objects as having heat, but they don't have heat. Heat is energy that is *moving* from a hot object to a cold object; neither object *has* heat. What objects have is *thermal energy*. Thermal energy is the total internal energy of a substance. Thermal energy is different from temperature. Temperature measures the average kinetic energy of all particles in a substance, not the total internal energy. Two objects can have the same temperature but different amounts of thermal energy. For example, a cup of coffee taken from a large coffee pot will be the same temperature as the coffee in the pot. But the coffee in the pot will have more thermal energy because there is more of it. The coffee in the cup will reach thermal equilibrium with its surroundings much sooner than the coffee in the pot. A very massive object that is cold has more thermal energy than a small object that is hot, even if the small object is very hot. Thermal energy depends on the total number of particles and on all other energies in the substance (e.g., bond energies and particle interaction energies). More particles (more mass) mean more thermal energy, regardless of the temperature of those particles. Thermal energy is measured in Joules, temperature is measured in degrees Fahrenheit or degrees Celsius (Westphal 2003).

Conduction is one of the mechanisms by which heat is transferred. For conduction to happen, the particles must be in contact with one another. The particles of the warm object must be touching the particles of the cool object (e.g., a metal spoon in a pot of hot soup). The particles in the warm object are moving faster than the particles in the cool object. At the point of contact, the faster moving particles of the warm object begin bumping into the slower moving particles in the cool object. As the cooler particles begin to warm, they move faster and bump into particles next to them. The heat flows from the fast-moving particles to the slow-moving particles until thermal equilibrium is reached (The Ohio State University 2016; Westphal 2003).

Conduction is why objects at the same temperature may feel warmer or colder than they are. Some types of materials conduct heat better than others. Metals are great conductors of heat. The particles in metals are freely moving, making it easy for heat to be transferred through a metal. In contrast, Styrofoam conducts heat much slower. The particles in Styrofoam are not as free moving as the particles in metals. It takes much longer for heat to move through Styrofoam. Think about what it feels like to touch metal that is at room temperature. It feels colder than other objects in the room. The objects in the room are at thermal equilibrium; they are all the same temperature. The metal feels colder because it is conducting heat away from your hand. You are warmer than the objects in the room, so the heat flows from you to the cooler object, which is the metal (The Ohio State University 2016; Westphal 2003).

## MISCONCEPTIONS

Common misconceptions about chemical reactions are discussed in Table 11.1.

### TABLE 11.1. COMMON MISCONCEPTIONS ABOUT TEMPERATURE, HEAT, AND THERMAL EQUILIBRIUM

| Common Misconception | Scientifically Accurate Concept |
|---|---|
| Heat is a substance. | Heat is energy that flows from one object to another. |
| Temperature is a property of a particular material or object. (Metal is naturally cooler than plastic.) | The temperature of an object is a measure of the average kinetic energy of the object. Objects may feel warmer or cooler because of their thermal conductivity. |
| The temperature of an object depends on its size. | The temperature depends on the average kinetic energy of the particles. |
| Hot and cold are different, rather than being opposite ends of a continuum. | Hot and cold are opposite ends of a continuous spectrum. The particles on the hot end of the spectrum are moving faster than the particles on the cold end of the spectrum. |
| Objects of different temperature that are in contact with each other or in contact with air at a different temperature do not necessarily move toward the same temperature. | Heat always moves from high temperatures to low temperatures. Regardless of the material or state of matter, heat will flow from hot to cold and thermal equilibrium will be reached. |
| Objects that readily become warm (conductors of heat) do not readily become cold. | An object that is a good conductor of heat will warm and cool quickly. |

*Source:* American Institute of Physics 2016.

## MATERIALS

- Aluminum cake pan (8 in.)
- Ice
- Mechanism for warming water (e.g., hot plate, immersion heater, electric tea kettle)
- Equal-sized samples of various materials such as aluminum, wood, glass, plastic, or ceramic (1 set per group)
- Stop watches (optional)
- Cups as close to the same size as possible

- 3 Styrofoam cups
- 3 glass cups
- 3 plastic cups
- 9 large containers to set cups in (approximately 32 oz.)
- 18 thermometers (nonmercury)
- Water at different temperatures:
    - » Ice cold
    - » Room temperature
    - » Hot
- Goggles (sanitized, indirectly vented chemical splash)
- Nonlatex gloves
- Aprons

## SUPPORTING DOCUMENTS

- "Measuring Temperature" data sheet
- "Transferring Thermal Energy" lab sheet
- "The Breakfast Bandit" prompt
- "Thermal Lunch Bag Challenge" prompt

## NONFICTION TEXTS

QR CODE 1

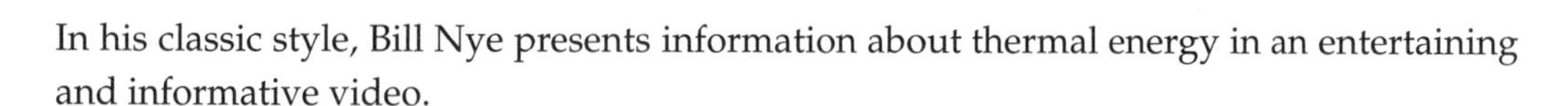

- "Bill Nye the Science Guy on Heat" by Disney Educational Productions, 2009, *http://bit.ly/1my7ARy.* (QR code 1)

In his classic style, Bill Nye presents information about thermal energy in an entertaining and informative video.

QR CODE 2

- "Heat Conduction" by CK–12 Foundation, *http://bit.ly/1ocsAPy;* Flesch-Kincaid reading level 9.0. (QR code 2)

This short article answers the question, "What is conduction?" The article combines everyday examples (e.g., a cookie sheet conducting thermal energy to the cookies) with accurate science.

QR CODE 3

- "How to Seal a Bag of Chips Without a Clip" by Howcast, 2008, *http://bit.ly/1WqYeTP.* (QR code 3)

This short video clip demonstrates how to cleverly seal a bag of chips with a few simple folds.

QR CODE 4

- "Thank Science for These 10 Awesome Hacks" by Thor Jensen, Tested, *http://bit.ly/1U7czpZ*. (QR code 4)

The stars from MythBusters present a set of slides that share the science behind 10 life hacks.

QR CODE 5

- "Thermal Conductors and Insulators" by CK–12, *http://bit.ly/1Pkkz*ka: Flesch-Kincaid reading level 6.4. (QR code 5)

This brief article discusses the difference between thermal conductors and insulators and provides examples of each.

## SAFETY CONSIDERATIONS

- Have students wear goggles, gloves, and aprons during all phases of the investigation, including during setup and cleanup.
- Review safety procedures for working with heated materials, and supervise students carefully as they conduct these investigations.
- Remind students to use caution when heating or working with hot liquids or solids, which can seriously burn skin.
- Remind students to use caution with heating devices such as Bunsen burners and hot plates. Tell them to never touch the heating device until there is a cool-down period to prevent burning the skin.
- Use only GFI-protected electrical outlets when working with hot plates or other electrical heating devices.
- If glass thermometers are used, tell students to handle them with care, because glass thermometers can break and puncture skin.
- Remind students to immediately wipe up any spilled water off the floor to prevent a slip or fall hazard.
- Have students wash their hands with soap and water after completing the investigation.

## LEARNING-CYCLE INQUIRY

### Engage

In this phase, students will learn about a "life hack" for thawing something.

1. Begin with a conversation about life hacks.
    a. Ask students what they know about the concept of life hacks. They are most likely familiar with the idea. If not, explain what a life hack is.

b. Show the "How to Seal a Bag of Chips Without a Clip" video (*http://bit.ly/1WqYeTP;* QR code 3). You may want to demonstrate the technique to show that it really works.

QR CODE 3

c. Share with students that some life hacks are actually based on science. Share one or two of the slides from "Thank Science for These 10 Awesome Hacks" (*http://bit.ly/1U7czpZ;* QR code 4).

QR CODE 4

2. Follow the chip-bag demonstration with another science-based life hack—a way to thaw or melt things quickly.
    a. Place an 8 in. aluminum cake pan in plain view. Show the class an ice cube, and ask them what they think will happen when you put the ice cube in the cake pan.
    b. Put the ice cube in the cake pan, and have students watch as it melts very quickly. To increase the drama, you might also leave an additional ice cube sitting in a plastic bowl next to the cake pan.
3. Ask students what they think might have happened. At this point, they are likely to share some scientifically inaccurate thinking. What they share can be helpful in determining what they already know about the thermal properties of materials.
4. Introduce the question, *How does an aluminum cake pan melt ice so fast?*

***Assess this phase:*** Assessment in this phase is limited to formative assessment. Student responses to questions about what they think happened may reveal common misconceptions about the thermal properties of matter.

## Explore

In this phase, students work from the macroscopic level to the molecular level as they investigate thermal energy. They will investigate the physical properties of various materials, quantitative and qualitative measures for temperature, transfer of thermal energy, thermal equilibrium, and conduction.

Students will need to have access to hot water in the Explore phase. Options for warming the water (if you do not have hot, running water in your classroom) include allowing individual student teams to heat their own water using a Bunsen burner, hot plate, or immersion heater or heating large quantities of water for the whole class using a slow cooker, electric skillet, electric teapot, or something similar.

### PART I: MELTING ICE

In this part of the Explore phase, students will develop a procedure and collect data to compare the rates at which ice melts when placed on a variety of materials (aluminum, wood, ceramic, plastic, and glass). At the end of this step, students will recognize that some property of the materials affects the rate at which ice melts.

1. Organize the class into groups of three to four students. Tell students that to confirm that ice really does melt faster on aluminum than on other materials, they will conduct an experiment to compare how long it takes ice to melt when it is placed on a variety of different materials. Before designing the experiment, groups should examine the materials and list the physical properties of each.
2. Give the class time to brainstorm what procedure they should use to fairly test the materials and what data they should collect. Share with them the materials and equipment that are available for their use.
3. Each group will now design a procedure for testing the rate at which ice melts when placed on various types of materials. Remind them that, at a minimum, the procedure must include the following:
    - Materials list
    - Steps
    - Tables for data collection
    - Plan for data analysis
4. At the very least, the steps of the procedure should include timing how long it takes for the ice cubes to melt. Groups could also stop the process after the ice cube that is melting fastest has completely melted. They could then measure either the volume of melted water or the mass of the remaining ice. Procedures should include a plan for recording and analyzing data. Groups should also think about how they will share their procedure and results as part of a class gallery walk. (*Note:* For an explanation of the gallery walk strategy, see page 342 in Appendix 2.) Guiding questions to ask as groups are designing their procedure include the following:
    - What are you doing to make sure you are conducting a fair test?
    - How will you know for sure which ice cubes melted fastest, slowest, or in between?
    - What variable are you measuring? (Possibilities include time, mass of remaining ice cube, and volume of melted water.)
    - How do you plan to analyze your data?
    - What type of graph will you use to represent your findings?
    - How will you share your procedure and results with the rest of the class?
5. Check each group's procedure before they begin. Have the materials samples and equipment readily available for students to use. Ask them to predict which ice cubes will melt the fastest.

6. Monitor students as they conduct their observations and collect data. Circulate around the room, asking guiding questions and providing support where needed. Monitor students' thinking as well as their behavior. Possible questions are as follows:

    - What are you noticing?
    - How do your results compare with what you thought might happen? (Students may have predicted that the ice would melt slower on the substance that feels the warmest.)

7. When all groups have completed their testing, conduct a gallery walk for groups to share their procedure and results with the class. Regardless of their procedure, each group should see the same results in terms of the rates of melting. (Ice placed on aluminum will melt fastest, and ice on wood will melt slowest.)

8. After the gallery walk, lead a class discussion about the results. Guide students toward the recognition that the physical properties of each material affect the rate at which the ice melts. Guiding questions could include the following:

    - How do the results of each group compare? (Students should have noticed that regardless of the procedure, the results are the same.)
    - Before conducting the experiment, what did you notice about the physical properties of each of the materials?
    - How might the physical properties of the materials be affecting the results?
    - What additional information is needed to answer our question, *How does an aluminum cake pan melt ice so fast?*

***Optional step:*** Some students may jump to the conclusion that ice melts faster on metal and shut down their thinking. If that seems to be the case, you should do a quick demonstration that will leave them scratching their heads. Put the aluminum cake pan used earlier and a galvanized zinc end cap (used in heating systems and available at home improvement stores) in view of all students. Put a piece of ice in both. The ice in the aluminum pan will melt quickly; and the ice in the zinc end cap will melt much slower. To keep the conditions as similar as possible, use an 8 in. zinc end cap. It is close enough in size and design that some students might mistake the end cap for a different type of cake pan.

### PART II: UNDERSTANDING TEMPERATURE, CONDUCTION, AND THERMAL EQUILIBRIUM

***Advance preparation:*** Set up three stations to demonstrate that conductors do not accurately measure temperature. Place three glasses of the same size (one each of glass, Styrofoam, and plastic) and three thermometers at each station. Put a container of ice water at station 1, room temperature water at station 2, and hot water at station 3.

After preparations are made, begin the Explore phase. The steps are as follows:

1. Direct students to set out their materials samples and place their palm on each one, one at a time. Ask the class what they notice. Students will notice that some are "colder" than others. Ask if they would be surprised to discover that they are all the same temperature. They will most likely be surprised by that information and may challenge the statement.
2. Explain that there are three stations in the room that provide evidence that items that are at the same temperature may actually feel warmer or colder. Before beginning, ask student volunteers to pour equal amounts of water (e.g., 200–250 ml) in each cup at each station and then place the thermometers in the cups. Give each student a copy of the "Measuring Temperature" data sheet.
3. Ask groups to move from station to station. At each station, groups will check the temperature of the water in each cup to confirm that they are the same. They will then touch the side of the cups to compare the perceived temperatures. The measured temperatures at each station should be the same for all of the cups (if the thermometers are properly calibrated). The estimated temperature determined by touching the cups will vary widely from cup to cup and station to station. The glass cup will feel the hottest or coldest, the Styrofoam cup will have little temperature variation, and the plastic cup will be somewhere in between.
4. After all groups have visited each station and completed their data collection sheets, discuss the results. The discussion will reveal two ways of thinking about temperature: (1) temperature as measured quantitatively with a thermometer and (2) temperature as measured qualitatively by touch. The following guiding questions may be helpful:

    - How did the temperatures measured with the thermometers compare between the cups at each station? (*Note:* Slight variations in temperature may be due to differences in the calibration or accuracy of the thermometers.)
    - How did the estimated temperatures compare to the measured temperatures?
    - How can you explain the differences between the results of the two measuring techniques?

    At this point, you may want to reiterate that temperature, as measured with a thermometer, is a measure of the average kinetic energy of a substance. If necessary, review what kinetic energy is. Determining temperature by touch is actually measuring the relative coldness or warmness of the materials. Explain that *temperature* is an example of a word we use in our daily life in a slightly different way than scientists use it in a scientific discussion.

    - What questions do you have? (*Note:* If nobody asks a question, wonder aloud why things at the same temperature can feel different.)

At this time, students may be wondering how things can feel as if they are at different temperatures when thermometers indicate that they are not. It may be hard for students to accept that the sense of touch is not a good indicator of temperature.

5. Ask students to read "Thermal Conductors and Insulators." As they read, they should think about how the text relates to the experiment they just completed. Before they read, students should create a double-column chart like the one in Figure 11.1. As they read, they should record their findings on the chart. Students should also note any vocabulary words that are new to them.

#### FIGURE 11.1. CONNECTING SELF TO TEXT FOR "THERMAL CONDUCTORS AND INSULATORS"

| What I learned from the article | Connections to what I learned in class |
|---|---|
| | |

6. After students have finished reading the article, relate the reading to their experimental results, using the following guiding questions:
   - Which of the cups is a better thermal insulator? Which is a better conductor? How do you know?
   - Which method of measuring temperature—by touch or with a thermometer—produces the truest measurement?
   - Why do cookie recipes tell you which temperature to bake the cookies at rather than saying something such as "bake in a warm oven"?
   - Why do metal objects feel colder than cloth objects?
7. Students will now explore the concept of thermal equilibrium. Distribute the "Transferring Energy" investigation sheet and the necessary materials. Students will conduct the experiment as written. As they work, circulate around the room, asking questions and providing support where needed.
8. When students have finished their investigation, discuss the results with the class. The following are several important points that should come out of the discussion and activities:
   - The temperature of the hot water decreased while the temperature of the cooler water increased.

- Once the temperature of the water in the two containers in a setup reached the same level, the temperature stopped changing (thermal equilibrium).
- The type of material the small cups were made of influenced the rate of temperature change. The Styrofoam cup warmed much slower.
- The thermal energy moved from the warm water to the cold water.

Consider using some of the following guiding questions to help students draw those conclusions:

- What did you observe?
- What happened to the temperature of the hot water (increase, decrease, stay the same)? The cold water?
- How did the cooling rate of the hot water in the two setups compare? How can you explain this?
- At what point did the hot water stop cooling? (*Answer:* When the temperature of the water in the large and small containers was the same.)

9. Before continuing, students need to define *thermal energy.* Ask them to refer to the notes they took while reading "Thermal Conductors and Insulators." Ask if anyone listed thermal energy as a vocabulary word that was new to them. Most likely, at least one student will have listed thermal energy. Provide a definition for thermal energy. *Thermal energy* is the total amount of energy in an object. Elaborate on the definition by explaining that objects have more than just the kinetic energy measured by a thermometer. They also have energy that holds atoms together in molecules and energy that holds molecules together. All of this energy combined is the thermal energy of a substance. Thermal energy cannot be measured with a thermometer because it involves more than kinetic energy. (This approach may seem a bit backwards, but thermal energy is typically difficult to define from clues in the texts that use the term.)
10. Now that a definition and explanation have been provided, draw a double-column chart like the one in Figure 11.2 where everyone can see it. Ask students to share any evidence they can think of from the text or in class activities that illustrates the meaning of thermal energy.

### FIGURE 11.2. DEFINING VOCABULARY WORDS TOGETHER

| Evidence from the text | Evidence from class activities |
|---|---|
| | |

11. Follow with questions about the direction in which the thermal energy flowed, including, "Which direction did the thermal energy move? From hot water to room temperature water, or from room temperature water to hot water?"
12. When you have anecdotal evidence that the students have grasped the concept of thermal energy and the flow of thermal energy from hot to cold, ask students to draw a diagram that illustrates the movement of the thermal energy. Ask students to share and discuss their diagrams.
13. After the discussion, explain the concept of thermal equilibrium in a process similar to the one used for thermal energy. This time, all of the evidence will come from class activities. Students should know that when objects of different temperatures are brought together, thermal energy flows from the warmer object to the colder object until both objects are the same temperature.

#### PART III: UNDERSTANDING MASS, THERMAL ENERGY, AND TRANSFER OF ENERGY

In this part of the Explore phase, students will determine that the more mass, the more kinetic energy required to raise the temperature

***Advance preparation:*** Before beginning the experiment, complete the steps that follow.

1. Heat enough water to submerge the objects that will be heated.
2. Put the objects to be heated in the hot-water bath. Allow sufficient time for the objects to reach thermal equilibrium before students retrieve them for their investigation. Have a thermometer ready to measure the temperature of the water just before the objects are removed.

After preparations are made, begin the experiment. The steps are as follows:

1. Share with students that they are going to gather data to determine if there is a relationship between the mass of an object and the amount of thermal energy it has.
2. Explain and demonstrate the experimental setup:
    a. Pour 200 ml of room-temperature water into a Styrofoam cup. Measure and record the temperature of the water. Record the temperature where everyone can see it.
    b. Measure and record the temperature of the water in the hot-water bath.
    c. Use tongs, or a slotted plastic spoon, to carefully remove one of the heated objects. Quickly put the object in the room-temperature water.
    d. Record the initial temperature of the water. Record the temperature again in 2 minutes. Continue recording the temperature until you get the same measurement twice.
    e. Remove the object from the cup and then dry and mass it.

3. Instruct students to pour equal volumes of room-temperature water into three separate Styrofoam cups. Then, tell them to add one object from the hot-water bath to each cup. All objects should be the same type of material but have different masses (e.g., 50 g mass, 100 g mass, and 150 g mass for 2, 4, and 6 heated pennies, respectively). Before they begin, students should set up a data table for recording their findings. After students have finished collecting data, they should graph their findings.
4. When all groups have finished, discuss their findings. Questions might include the following:

   - What happened to the temperature of the water in the cups when the heated objects were added? (*Answer:* The temperature of the water increased.)
   - How did the change in water temperature compare for the different-size objects? (*Answer:* The more massive the object, the greater the temperature increase.)
   - What was the starting temperature of the objects? (*Note:* This question could stump the students. The temperature of the objects is the same as the temperature of the hot-water bath because of thermal equilibrium. If students are having difficulty relating this question to thermal equilibrium, ask them if thinking about thermal equilibrium would help them determine the temperature of the objects.)

5. Ask students to predict what would happen if the test was repeated with different objects. Repeat the process with different objects. If desired, repeat the process a third time with a third substance. Repeating the test allows students to look for patterns and trends that can be used to generalize their results.
6. Discuss all results and ask students to predict what would happen if you added cool items to hot water (e.g., ice). Challenge students to devise a fair test and conduct the experiment. Follow the experiments with a discussion of the results.
7. After all student tests have been conducted and after results have been discussed, ask students to read "Heat Conduction" to learn how conduction works at the molecular level. As they read, they should connect what they are reading to the investigations they have been conducting, just as they did when they read "Thermal Conductors and Insulators."
8. Follow this by showing "Bill Nye the Science Guy on Heat." After showing the video, ask students to connect information in the video to what they have learned in class.
9. Close the Explore phase by asking students, *How does an aluminum cake pan melt ice so fast?*

*Assess this phase:* The Explore phase has many opportunities for formative assessment. Anecdotal evidence gathered from discussions, gallery walk presentations, and observations all provide valuable information about students' progress. Questions students ask you and one another can also reveal their level of understanding at the time. The way they analyze the data they collect in each of the lab activities can also be used as a formative assessment. The "Science Process Skills Rubric" found in Appendix 1 could be used as a summative assessment of students' science process skills throughout the unit.

## Explain

Students will apply what they know about thermal energy to a fictional scenario: the mysterious case of the Breakfast Bandits. Students will read "The Breakfast Bandit" and respond to the RAFT prompt. Distribute a copy of the prompt to each student. Students will work independently as they respond to the prompt.

*Assess this phase:* Student responses to the writing prompt are used as summative assessments in this phase of the unit.

## Expand

In the Expand phase, students will apply what they know to design a thermal lunch bag for keeping food warm or cold for several hours.

1. Collect a variety of materials for students to use in their design (e.g., plastic grocery bags, clean towels or rags that can be cut up if necessary, polyester fiber fill, aluminum foil, large pieces of plastic, bubble wrap, packing peanuts, newspaper, reusable shopping bags, small boxes). Be creative with the materials. Make sure some are good insulators and some are good conductors. Students will also need small containers with lids that can be used as the hot or cold "lunch." When they are testing their lunch bags, they will fill the container with hot or cold water and put it in their thermal lunch bag.
2. Students will work in teams to design the thermal lunch bag. Give each team a copy of the *Lunch Bag Challenge* sheet. Instruct students as to what materials they can use in their construction.
3. After students have finished the first round of testing, have them share their results with the class. Groups should then modify and retest their design. After all testing is finished and the results have been shared with the class, ask students to complete a 3-2-1 exit slip before they leave class. They should list three ways they applied what they have learned about thermal energy to their lunch bag design, two ways they can improve on their design, and one problem they are having a difficult time solving.

*Assess this phase:* The 3-2-1 exit slip can be used for formative assessment.

## REFERENCES

American Institute of Physics. 2016. Children's misconceptions about science. *http://amasci.com/miscon/opphys.html.*

National Governors Association Center for Best Practices and Council of Chief State School Officers (NGAC and CCSSO). 2010. *Common core state standards.* Washington, DC: NGAC and CCSSO.

Nave, C. R. 2016. Heat and thermodynamics. HyperPhysics. *http://hyperphysics.phy-astr.gsu.edu/hbase/thermo/heat.html.*

NGSS Lead States. 2013. *Next Generation Science Standards: For states, by states.* Washington, DC: National Academies Press. *www.nextgenscience.org/next-generation-science-standards.*

The Ohio State University. 2016. Chapter 4: Transfer of thermal energy. *www.physics.ohio-state.edu/p670/textbook/Chap_4.pdf.*

The Physics Classroom. 2016. Introduction to Thermal Physics. *www.physicsclassroom.com/class/thermalP/Lesson-1/Introduction.*

Westphal, P. 2003. Temperature, heating, and thermal energy—Sorting things out! *http://modeling.asu.edu/modeling/Heating-Temp_PWestphal03.doc.*

Name______________________________ Date____________________

# Measuring Temperature

## BACKGROUND

When we think about temperature in our day-to-day life, we think about the hotness or coldness of something. This is a descriptive measure of temperature. We often determine something's hotness or coldness by touching it. We think of temperature differently in science. In science, we think of temperature as a measure of the average kinetic energy of a substance. A high temperature means that the particles in the substance are moving fast. They have a high kinetic energy. A low temperature means the particles are moving slowly. They have a low kinetic energy. When we use a thermometer to measure temperature, we are really measuring how fast or slow the particles in a substance are moving.

In this experiment, you are going to measure temperature by touch and with a thermometer.

## PROCEDURE

1. At each station, use the thermometer to measure the temperature of the water in each cup. Record the information in Table 1.
2. Touch the side of each cup, one at a time, at each station. Rank the cups from warmest to coldest in Table 2.

### TABLE 1. MEASURING TEMPERATURE WITH A THERMOMETER

| Type of Cup | Station 1 | Station 2 | Station 3 |
|---|---|---|---|
| Plastic | | | |
| Glass | | | |
| Styrofoam | | | |

### TABLE 2. ESTIMATING TEMPERATURE BY TOUCH

| Temperature | Station 1 | Station 2 | Station 3 |
|---|---|---|---|
| Warmest | | | |
| In between | | | |
| Coldest | | | |

Which method of measuring temperature is the most reliable? Support your answer with evidence.

Name_______________________________________________ Date___________________________

# Transferring Thermal Energy

In this part of the investigation, you will discover what happens when a small container of room-temperature water is set in a larger container of hot water.

## MATERIALS

- Large beaker or plastic bowl (The beaker or bowl should be about twice the size of the small cups.)
- Small Styrofoam cup
- Small beaker or cup (about the same size as the Styrofoam cup)
- Hot water (approximately 500 ml)
- Room-temperature water (approximately 200 ml)
- 110 ml graduated cylinder (if not available, use measuring cups)
- 4 thermometers (nonmercury)
- Goggles
- Nonlatex gloves
- Aprons

## SAFETY REMINDERS

- Wear goggles, gloves, and aprons during all phases of the investigation, including during setup and cleanup.
- Follow your teacher's directions while working with heated materials.
- Be careful with devices such as Bunsen burners and hot plates. Never touch the devices unless they have had time to cool down.
- Be careful with glass thermometers.
- Immediately wipe up any spilled water off the floor.
- Wash your hands with soap and water after completing the investigation.

## PROCEDURE

1. Before beginning, you will need to determine how much water you will need to put in the large beakers and how you will keep the Styrofoam cup from floating.
    a. Follow these steps to determine how much water to put in the large beaker:
        i. Remember that you need enough water in the large beaker so that when you set the smaller cups inside, the water in the large beaker comes about ¾ of the way up on the side of the small beaker.
        ii. Set the small beaker inside the large beaker to find out how much water you need in the large beaker. Then, pour water into the large beaker (around the

Name________________________________ Date__________________

## Transferring Thermal Energy (*continued*)

outside of the small beaker) until the water in the large beaker is ¾ of the way up the side of the small beaker.

iii. Remove the small beaker and pour the water from the large beaker into a graduated cylinder to measure it. This measurement is how much hot water you will need to put in the large beaker when you begin.

b. Follow these steps to decide how to weight the Styrofoam cup down:

i. Remember that you need to weigh the Styrofoam cup down by setting something on top of it or else the cup will float and tip over in the water.

ii. Put the amount of water in the large beaker that you will be using (determined in step a). Now, put about 100 ml of water in the Styrofoam cup. You'll notice that the cup floats.

iii. Place something on top of the Styrofoam cup (e.g., a small notebook) to hold it in place. Make sure you can still put a thermometer into the Styrofoam cup.

2. Pour equal amounts (75–100 ml) of room-temperature water into the Styrofoam cup and the other small cup.
3. Pour equal amounts of hot water (determined earlier) into the large beakers.
4. Carefully set the cups of room-temperature water into the beakers of hot water. Make sure that none of the hot water flows into the small cups. Be especially careful with the Styrofoam cup because it will float. Hold it in place by the rim, or weight it down by setting something on top of it. *Do not weigh it down by putting something into the cup!*
5. As soon as the small cups have been put inside the larger beakers, put thermometers into all four cups.
6. Quickly measure the temperature of the water in all four cups. Record the temperature in your data table.
7. Measure the temperature of all four cups again in 2 minutes.
8. Continue measuring the temperature every 2 minutes until you get the same temperature readings twice in a row. Once you get the same temperature reading twice in a row, you can stop measuring the temperature in that cup.
9. Graph your results in a line graph.

Name________________________________________ Date______________________

### Transferring Thermal Energy *(continued)*

#### TEMPERATURE DATA TABLE

(Add additional rows, if needed.)

| Time (min.) | Example | Large Beaker With Small Cup (Setup A) | | Large Beaker With Styrofoam Cup (Setup B) | |
|---|---|---|---|---|---|
| | | Large Beaker | Small Cup | Large Beaker | Styrofoam Cup |
| 0 | 55°F | | | | |
| 2 | 51°F | | | | |
| 4 | 46°F | | | | |
| 6 | 41°F | | | | |
| 8 | 36°F | | | | |
| 10 | 36°F | | | | |
| 12 | | | | | |
| 14 | | | | | |
| 16 | | | | | |

What did you notice about the changes in temperature for each of the containers of water?

How did the temperatures of the large and small cup in setup A compare at the beginning (Time 0)?

Name_______________________________________________ Date________________________

**Transferring Thermal Energy** (*continued*)

How did they compare at the end?

How did the rate of temperature change for the cups in setup A compare to the temperature changes in setup B?

What conclusions can you draw about what happens when a small container of room-temperature water is set in a larger container of hot water?

Name_______________________________ Date___________________

# The Breakfast Bandit

Recently, there has been a string of home invasions in which the perpetrator enters a residence in the morning and takes one serving of the warm breakfast that has been prepared for the family. The perpetrator, dubbed the Breakfast Bandit, has a strange habit of leaving notes next to the bowls. The notes are always the same. One says, "too hot," another "too cold," and in the place where the missing serving sat, the note says, "just right." For example, in Home A, the breakfast servings were set out in identical bowls. Each bowl contained a different amount of food. The note next to the bowl with the largest serving said, "too hot." The note next to the smallest serving said, "too cold." The note where the bowl that had an average-size serving was sitting before it was taken said, "just right." In Home B, each bowl had the same serving size, but the bowls were different. The note next to the metal bowl said, "too cold." The note next to the insulated bowl said, "too hot." The note where the plastic bowl had been sitting before it was taken said, "just right."

You are the **science writer** at the local newspaper. You recently told your editor that you thought the notes had something to do with the **temperature of the food** and that the Breakfast Bandit was taking the food that was just the right temperature. The **local residents** need answers! Your editor has asked you to write an **article** that explains why the Breakfast Bandit is leaving these odd notes.

Name_______________________________ Date_________________

# Thermal Lunch Bag Challenge

The school cafeteria is in terrible shape! It is outdated, and some of the equipment no longer works correctly. The school is going to shut the cafeteria down for six weeks for renovations. That means you will have to pack your lunch for the next six weeks! Knowing that students will not look forward to packing their lunch, the principal has decided to challenge students to think outside the box (the lunch box, that is) about how they will transport their lunch to school. He has challenged the entire middle school to design lunch bags that they can use to bring healthy and delicious hot or cold lunches to school.

The principal's challenge is to design an inexpensive thermal lunch bag that can be reused multiple times. The bags have to be able to keep food warm or cold for approximately three hours. The design specifications state that the thermal lunch bag must be

- large enough to hold one covered dinner tray (approximately 4.5 × 8.25 × 2.25 in.),
- light enough that it could easily be carried with a complete meal inside (Remember that you will also be carrying your book bag!), and
- capable of keeping food from cooling or heating at a rate greater than 10°F per hour.

Your challenge is to work with your group to design, test, and present your findings. Groups have one day to design the lunch bag, one day to test it, and one day to present their findings to classmates. All groups *must* collect data and include them in their presentation. You and your group should also be able to explain the science behind the design.

Good luck!

# Chapter 12
# Landfill Recovery

## OVERVIEW

In this chapter, students gather evidence by completing hands-on activities and reading texts to answer the question, *How can we reduce the volume of landfill waste?* Students learn about the volume of waste that ends up in landfills every year, investigate the feasibility of landfill mining, engage in a landfill mining simulation, and design a product made entirely of reused plastics. Students also construct an argument about the pros and cons of landfill mining using evidence they collect from a variety for sources.

This unit assumes that students know and can demonstrate that some of Earth's natural resources are limited and that recycling extends the life of nonrenewable natural resources. Additionally, this unit contributes to attainment of *Next Generation Science Standards* performance expectation ESS3-3: "Apply scientific principles to design a method for monitoring and minimizing a human impact on the environment" (NGSS Lead States 2013).

## OBJECTIVES

- Recognize that interventions can slow or stop the negative impact of human activities on the environment.
- Identify strategies for reusing natural resources.
- Investigate ways to reduce municipal solid waste.
- Draw conclusions that are based on evidence from multiple sources.
- Engage in a scientific argument.

## STANDARDS ALIGNMENT

### *Next Generation Science Standards* (NGSS Lead States 2013)

#### ESS3.C: HUMAN IMPACTS ON EARTH SYSTEMS

- Human activities have significantly altered the biosphere, sometimes damaging or destroying natural habitats and causing the extinction of other species. But changes to Earth's environments can have different impacts (negative and positive) for different living things. (MS-ESS3-3)
- Typically, as human populations and per-capita consumption of natural resources increase, so do the negative impacts on Earth unless the activities and technologies involved are engineered otherwise. (MS-ESS3-3 and MS-ESS3-4)

### *Common Core State Standards, English Language Arts* (NGAC and CCSSO 2010)

#### GRADES 6–12 LITERACY IN HISTORY/SOCIAL STUDIES, SCIENCE, AND TECHNICAL SUBJECTS

- CCSS.ELA-LITERACY.WHST.6-8.1: Write arguments focused on discipline-specific content.
- CCSS.ELA-LITERACY.WHST.6-8.2: Write informative/explanatory texts, including the narration of historical events, scientific procedures/experiments, or technical processes.
- CCSS.ELA-LITERACY.WHST.6-8.9: Draw evidence from informational texts to support analysis, reflection, and research.

## TIME FRAME

- Seven to eight 45-minute class periods

## SCIENTIFIC BACKGROUND INFORMATION

Humans have been using Earth's resources for millennia. From early humans' use of flint for tools to our present-day use of technologically advanced materials, humans have recognized that using Earth's resources enhances our survival and quality of life. As humans advanced to form civilizations and use agricultural practices, the use of natural resources increased. It increased again when humans learned how to extract natural resources from the ground. When we humans learned how to manipulate, alter, and produce marketable products, our use of natural resources increased to extraordinary levels. As we began to use natural resources at increased rates, we also increased the amount of waste produced. Production and disposal of marketable items generates significant amounts of waste, and all of that waste has to go somewhere. In modern history, humans have moved waste from the view of the public eye to landfills, where it is buried and never to be seen again.

In 2013, Americans generated 254.1 million tons of trash (National Geographic 2016). That's well over 508 billion pounds of trash! Between 1960 and 2000, the amount of trash generated each year increased from 88.1 million tons to 243.5 million tons. Since 2000, the rate of increase has leveled off for the most part. But, overall, it is still creeping upward. See Table 12.3 (p. 245) to discover what humans are throwing away.

Recycling has helped reduce the amount of trash that makes its way into landfills, and recycling has increased over the years. In 1960, people recycled 6.4% of their trash; in 2013, people recycled 34.3% of their trash (EPA 2016). Increasing the rate of recycling is a necessary, but insufficient step toward reducing the amount of trash and nonrenewable natural resources that ends up in landfills. Although recycling has helped reduce the amount of trash going into landfills, the Environmental Protection Agency has stated that reducing the amount of trash generated and reusing items rather than disposing of them are the most effective ways to reduce solid waste (EPA 2015).

Mining landfills to recover valuable natural resources is an option that has been gaining traction in recent years. Some of the benefits of landfill mining, beyond the recovery of natural resources, include extending landfill capacity, lowering landfill operating costs, producing energy by incinerating materials that cannot be recovered, reducing landfill closure costs, retrofitting liners to repair leaks and tears, removing hazardous wastes, and generating income from recovered materials (EPA 1997). Drawbacks include managing hazardous materials, release of landfill gases and odors, subsidence or landfill collapse, and increased wear on excavation equipment (EPA 1997).

Several examples of successful landfill mining in the United States and abroad suggest that landfill mining may be part of the solution to our dwindling natural resources. The anticipated recovery rates of some resources, according to available information, is 85–95% for soil, 70–90% for ferrous metals, and 50–75% for plastics. The purity of the recovered materials ranges from 90% to 95% for soil, from 80% to 95% for ferrous metals, and from 70% to 90% for plastics (Environmental Alternatives 2016). A Belgian waste management company presently mining a landfill near Brussels expects to reclaim 45% of the materials; the rest of the materials will be converted to electricity (Vijayaraghavan 2011).

Landfill mining is not without drawbacks, however. Technology to mine landfills efficiently and safely has not yet been developed. Using current technology requires multiple steps to excavate, separate, and clean the resources. An important economic and environmental consideration is that recovering some of the materials consumes more energy than would be used to make new materials. Additionally, the cost to recycle the reclaimed materials costs much more than the market value of the resources. For example, in 2009 it cost $4,000 to recycling 1 ton of plastic bags, but the resulting product had a value of only $32 (Clean Air Council 2016).

Each of these options, recycling, reducing waste, reusing items, and landfill mining, offer some hope for extending the life of natural resources. Alone, none of the options is sufficient to address the problem of diminishing natural resources. But in combination, the various options could make a meaningful difference.

## MISCONCEPTIONS

Misconceptions about the availability of natural resources are based on fundamental misunderstandings about core Earth science principles. Most notably, some people think that humans depend on Earth for resources and that the activities of humans significantly alter Earth. Several of these misconceptions that are most closely related to this unit appear in Table 12.1 (p. 240).

**TABLE 12.1. COMMON MISCONCEPTIONS ABOUT NATURAL RESOURCES AND THE IMPACT OF HUMAN ACTIVITIES**

| Common Misconception | Scientifically Accurate Concept |
|---|---|
| Earth's resources are not finite. There is an endless supply of water, petroleum, and mineral resources. All we have to do is explore to find them. | Earth's resources are finite, and we have a very good idea about where we can find and harvest existing resources. Earth scientists also know approximately how abundant or lacking various Earth materials are. |
| • "Man-made" materials do not come from mineral resources.<br>• Few products we use everyday have anything to do with taking rocks and minerals from the ground. | All raw materials come from Earth. Products that are "man-made" (or human-made) are the result of processing raw materials through manufacturing. The plastics we use come from crude oil; all metals originate in ores mined from Earth; building materials, such as bricks, granite counter tops, and plasterboard, come from Earth materials. |
| Earth and its systems are too big to be affected by human actions. | Earth is a very large system, but scientists have accumulated overwhelming evidence that human activities alter Earth in irreparable ways. Earth's resources are limited and, in some cases, dwindling rapidly. Resources such as safe, potable water are difficult to access in some parts of the world. Human population now exceeds 7 billion people. As the population continues to grow, the strain on resources such as potable water will increase. Humans can and do affect Earth systems. |
| Technological fixes will save us from ruining our planetary environment. | Technology and science have limitations. It is true that technology and science can reverse or remedy some of the environmental damage that humans have caused. However, so far no one has developed technology to solve all of our environmental problems. It is unlikely that our current scientific knowledge and technology will ever generate more of the nonrenewable resources we take from Earth. Moreover, these advancements cannot solve current problems such as nuclear waste spills or toxic contamination such as that found at some Superfund sites. |
| Earth is both an endless supply of resources and a limitless sink for the waste products of our society. | We used to believe the adage that "the solution to pollution is dilution." We now know that such thinking not only is wrong but also threatens the health of Earth and the organisms that live on it. Even small concentrations or low doses of some toxic compounds can render soil or water unsafe for decades to come. |

*Source:* The Math and Science Partnership Network 2016.

## NONFICTION TEXTS

*Heroes of the Environment: True Stories of People Who Are Helping to Protect Our Planet* by Harriet Rohmer (San Francisco, CA: Chronicle Books, 2009); Flesch-Kincaid reading level 7.4, published Lexile level 1070L.

The stories of 12 people who are taking inspiring actions to protect the planet are shared in this engaging book. The stories range from removing industrial pollution from a river to protecting sea turtles and whales. The individuals portrayed are just as diverse as their actions.

*One Plastic Bag: Isatou Ceesay and the Recycling Women of the Gambia* by Miranda Paul (Minneapolis, MN: Millbrook Press, 2015); Flesch-Kincaid reading level 2.9, published Lexile level 480L.

This children's picture book carries a powerful message about the waste we generate, the impact it has, and the ways we might reduce it.

## MATERIALS

### For the Landfill Mixture

Note that you can substitute other materials. Choose materials according to the separation techniques needed. If working with salt or coffee grounds, be careful of items getting wet.

- Dry pasta
- Iron filings (see safety data sheet information in Appendix 4)
- Shredded paper
- Crayon shavings (made with a pencil sharpener)
- Mini-marshmallows
- Sand

### For the Rest of the Investigation

- Magnets covered with plastic wrap
- Balloons (nonlatex)
- Tub or large bowl of water
- Pans or bowls for separating with water
- Beakers, cups, or bowls for the separated materials

- Heat source and container for melting (1 per group or 1 for the class, if done as a demonstration)
- Crayon mold
- String
- Bucket or gallon milk jug with an opening cut in the top that is large enough to add weight
- Masses, washers, sand, or water for testing
- Sufficient quantity of plastic grocery bags, empty water bottles, and K-cups (if you choose to use them) for the design challenge
- General supplies, such as scissors, tape, and glue
- Goggles (sanitized, indirectly vented chemical splash)
- Nonlatex gloves
- Aprons

## SUPPORTING DOCUMENTS

- "Turning Waste Into Good Business and Good Jobs" graphic organizer
- "One Year of Solid Waste" cards
- "Municipal Solid Waste" cut outs
- "Landfill Mining: Brilliant Idea or Wishful Thinking?" essay and graphic organizer
- "Based on the Evidence … " graphic organizer
- "Recovering Resources" planning and data sheet
- "Strength Test Data Sheet"
- "Recommendation to Style Barons" graphic organizer
- "Individual Recommendation to Style Barons" graphic organizer

## SAFETY CONSIDERATIONS

- Have students wear goggles, gloves, and aprons during all phases of the investigation, including during setup and cleanup.
- Review safety procedures for dealing with heated materials, and supervise students carefully as they conduct the investigations.
- Review safety information in safety data sheets for chemicals (e.g., iron filings and plaster of paris) with students (see Appendix 4).
- Remind students to use caution in working with iron filings; they can be sharp and puncture skin.
- Tell students not to breathe in the dust from these filings.

- Make sure students have appropriate procedures for cleanup and disposal of iron filings.
- Remind students to use caution when heating or working with hot liquids or solids, which can seriously burn skin.
- Remind students to use caution with heating devices such as Bunsen burners and hot plates. Tell them not to touch the heating device until after it has cooled down.
- Use only GFI protected electrical outlets when working with hot plates or other electrical heating devices.
- Remind students to immediately wipe up any spilled water off the floor to prevent slip and fall hazards.
- Remind students not to eat any food used in lab investigations.
- Have students wash their hands with soap and water after completing the investigation.

## LEARNING-CYCLE INQUIRY

### Engage

In the Engage phase, students are introduced to the idea of reusing materials that are headed for a landfill. Students independently read "Turning Waste Into Good Business and Good Jobs," which is a chapter from *Heroes of the Environment.* As students read the chapter, they will discover that discarded items can be valuable and present opportunities for reuse.

***Advance preparation:*** Make one copy of "Turning Waste Into Good Business and Good Jobs" for each student. Make one copy of the "One Year of Solid Waste" cards per group of three to four students. Cut the cards apart, and shuffle them before distributing a set to each group.

After preparations are made, begin the Engage phase. The steps are as follows:

1. Ask students to share something they do or something they know of that helps the planet. Then, introduce the book *Heroes of the Environment.* Describe the types of things the heroes of the book are doing to help keep the planet healthy.
2. Distribute a copy of "Turning Waste Into Good Business and Good Jobs" (Chapter 3 of *Heroes of the Environment*) and the accompanying graphic organizer. Tell students that they are now going to read about Omar Freilla from the Bronx in New York. As they read, students should be looking for the message of Omar's story. After they have finished reading, they should write what they think the message of Omar's story is in the appropriate space on the graphic organizer. Students should then add evidence that supports their thinking about the message of Omar's story.
3. After students have finished the reading, ask them to turn and talk to a neighbor about what they think Omar's message is. Then, ask several students to share what

they took away from the story. Make sure that the class has identified Omar's message: If something has a use, it is not waste.

4. Have students work in small groups to explore the amount of solid waste generated in the United States in one year.
    a. Begin by providing each group of students with one set of the "One Year of Solid Waste" cards.
    b. Ask students to work together to match the statistic card with the correct item.
    c. After 5–10 minutes, ask each group to share its results. As groups share, ask them to explain why they matched them up as they did.
    d. After all groups have shared, ask each group to review its initial matches and make changes that reflect the group's current thinking.
    e. When students are finished, reveal the correct matches shown in Table 12.2. Reinforce that these statistics represent one year of solid waste.

## TABLE 12.2. ONE YEAR OF SOLID WASTE CARDS

| | |
|---|---|
| Amount of trash generated by Americans in 2013 | 254 million tons of trash |
| Pounds of trash disposal per person | 4.6 pounds |
| Percentage of trash from residences | 65% |
| Percentage of trash buried in landfills | 55% |
| Percentage of trash from schools and commercial locations | 35% |
| Percentage of trash recycled | 33% |
| Percentage of trash incinerated | 12.5% |
| Number of solid-waste industry employees | 368,000 |
| Number of vehicles used to move trash to landfills | 148,000 |
| Number of communities with curbside recycling | 8,660 |
| Number of landfills | 1,754 |
| Number of recycled-materials sorting centers | 545 |
| Number of incinerators | 87 |
| Solid-waste industry annual revenue | $47 billion |

*Source:* National Geographic 2016.

5. Engage students in a discussion about what they think they would find if they dug deep into a landfill. Students are likely to offer unspecific responses such as garbage,

trash, rotten stuff, and so on. Those answers are acceptable at this point. List the students' responses where all students can see them.

6. Students will now find out what is in landfills. Specifically, they will learn what percentage of municipal solid waste is paper, glass, metals, and so forth. For this activity, students will work in their groups.
   a. Give each group a copy of the "Municipal Solid Waste" page and a pair of scissors.
   b. Instruct students to cut categories at the bottom of the page apart and rank them, from highest to lowest, according to which ones they think make up the most or least amount of waste in a landfill.
   c. After students have ranked the categories of waste, they should match them to the appropriate section of the pie chart on the basis of their ranking.
   d. Ask groups to share where they placed each category on the pie chart and explain their reasoning.
   e. Compare and discuss groups' placements of the cards on the pie chart. After the discussion, ask groups to make any changes they want. This time, they should tape the cards in place.
   f. Reveal the correct percentages, using the information in Table 12.3.
7. Finish by introducing the question, *How can we reduce the volume of landfill waste?*

**TABLE 12.3. MUNICIPAL SOLID WASTE**

| Category | Percent |
|---|---|
| Paper | 27.0 |
| Food | 14.6 |
| Yard trimmings | 13.5 |
| Plastics | 12.8 |
| Metals | 9.1 |
| Rubber, leather, and textiles | 9.0 |
| Wood | 6.2 |
| Glass | 4.5 |
| Other | 3.3 |

*Source:* EPA 2016.

***Assess this phase:*** Assessment of this phase is formative. Students are likely to reveal misconceptions about solid waste as they complete the "One Year of Solid Waste" and "Municipal Solid Waste" matching activities.

## Explore

In this phase, students will continue working in groups to investigate the feasibility of mining landfills as a means to recover some discarded resources. They will also participate in a simulation in which they recover and test a nonrenewable resource.

### PART I: MINING LANDFILLS

1. Provide each student with a copy of "Landfill Mining: Brilliant Idea or Wishful Thinking?" Instruct students to independently read the essay and complete the graphic organizer before returning to their groups.
2. After students return to their groups, distribute the "Based on the Evidence …" graphic organizer. Instruct students to discuss which company they think Style Barons should hire. Remind them that their decision must be based on evidence. Circulate around the room, asking guiding questions such as the following:
   - What evidence leads you to think Style Barons should choose this company?
   - How could you compare the materials available through traditional mining with materials recovered through landfill mining?
   - What additional evidence would help you make your decision?

### PART II: COMPARING MATERIALS

***Advance preparation:*** Collect old crayons for shaving. Remove the paper wrapping from the crayons, then use a manual pencil sharpener to make the shavings. Save some crayons to use when demonstrating how to melt the crayons and pour them into the mold. Combine the materials to make the landfill mixture. Bag and label the mixture. Make plaster of paris crayon molds.

After preparations are made, begin Part II of the Explore phase. The steps are as follows:

#### Separation

1. Before beginning, gather up separation tools and general materials. Put them in a central location where groups can easily access them during the separation activity.
2. Set the activity up with the following scenario:

   *Each company has sent us some materials to test. Earth Materials R Us sent a pure sample of one of the materials they have available. Reclaiming the Past sent us a sample of items mined from their landfill. These items have gone through a washing process, so they are safe for us to handle. We have to sort the materials to recover what Style Barons needs. After we recover the materials, we must make a sample similar to the one Earth Materials R Us sent. I am expecting a mold from them to arrive any day.*

3. Give each group a bag of the landfill mixture. Provide each student with a copy of the "Recovering Resources" planning and data sheet.
4. You are playing the role of the lab manager. As groups work on designing and carrying out their separation process, circulate around the room to check on their designs and ask guiding questions.

5. After students have separated their materials, they will take turns melting the materials (crayon shavings) and pouring them into their molds.

### Testing

1. Before you begin testing, reveal that the molds have arrived. Show the students the molds and demonstrate how they will be used.
    a. Show the students how to melt the crayons. Set up a single melting station that can be closely monitored. The crayons melt easily in a microwave, a candy melting pot, or in glass beakers that are in a hot-water bath. (An electric skillet works great for a hot-water bath.) It is helpful to melt the crayons in a container with a pour spout (e.g., a beaker or measuring cup). Doing so makes it much easier for students to accurately pour the melted crayons into the mold. While the crayons are melting, lubricate the mold by rubbing a drop of dishwashing detergent inside.
    b. Model how to handle the beaker with the melted crayons and how to pour the melted crayons into the mold. The melted crayons will rapidly start to harden again, so students should bring their molds to the melting station when they are ready to make their crayons. Once the melted crayons have been poured into the molds, groups can move them to a safe place for setting. Although they will set quickly, let the molds continue to set overnight so they are completely cooled before testing.
    c. Demonstrate how to remove the crayons from the molds by flipping the mold over. The crayon should fall out.
2. It is now time to test the strength of a pure crayon (the pure sample from Earth Materials R Us) and the one made from the reclaimed crayons. To contain the excitement and monitor testing, groups should test their crayons one at a time.
    a. Before beginning, separate two desks or tables by enough space so that the crayons span the opening with each end of the crayons securely on the desks. Give each student or group a "Strength Test Data Sheet."
    b. Mass the crayon to be tested.
    c. Slide the crayon to be tested through the handle of the loading device.
    d. Position the crayon so that it spans the gap in the desks.
    e. Slowly add mass (e.g., sand or marbles) to the loading device. Stop adding mass when the crayon breaks.
    f. Measure the amount of mass the crayon held before breaking.
3. Have students complete their data sheets and draw conclusions about which material is the better choice for Style Barons.

*Assess this phase:* The Explore phase for this learning-cycle inquiry can be formatively assessed. As students gather evidence from "Landfill Mining: Brilliant Idea or Wishful Thinking?" and from separating and testing the recovered materials (crayon shavings), they should be processing their thinking about the volume of waste that ends up in landfills and the possibility of solutions such as reducing waste and reclaiming resources found in landfills. Monitor their progress by listening carefully to their discussions as they work and by asking probing questions.

## Explain

In this phase, students will use the data collected in the Explore phase to construct a scientific argument in response to the prompt, *Which company do you recommend that Style Barons use?*

1. Provide each student with a "Recommendation to Style Barons" graphic organizer.
2. Instruct students to work with their groups to review and list their evidence on the "Recommendation to Style Barons" sheet.
3. As groups work, walk around the room to ask guiding questions and provide assistance when needed.
4. When groups have finished their discussions, give each student an "Individual Recommendation to Style Barons" sheet. Tell students to work independently to construct an argument that supports their recommendation of either Earth Materials R Us or Reclaiming the Past.
5. Circulate around the room as students work independently, providing assistance and support when needed.

*Assess this phase:* In the Explain phase, each student's "Individual Recommendation to Style Barons" serves as a summative assessment.

## Expand

In this phase, students will design a product that is made entirely of recycled plastics.

*Advance preparation:* Collect plastic shopping bags, water bottles, and K-cups for students to use in their design challenge. Be aware that used coffee grounds, tea leaves, etc. will still be in the K-cups.

After preparations are made, begin the Expand phase. The steps are as follows:

1. Project the infographic "Sneaky Plastic Waste We're all Producing" shown in Figure 12.1. Discuss the volume of plastics going into landfills from just these three products. Talk about personal and observed use of these plastics.

**FIGURE 12.1. SNEAKY PLASTIC WASTE WE'RE ALL PRODUCING**

One trillion are used worldwide

Less than 5% end up recycled

Takes 20–1,000 years to degrade completely

In 2013, Keurig cups would wrap around the equator 10.5 times

100% of this #7 plastic contains synthetic estrogen

Made of a type of plastic that doesn't recycle in most areas

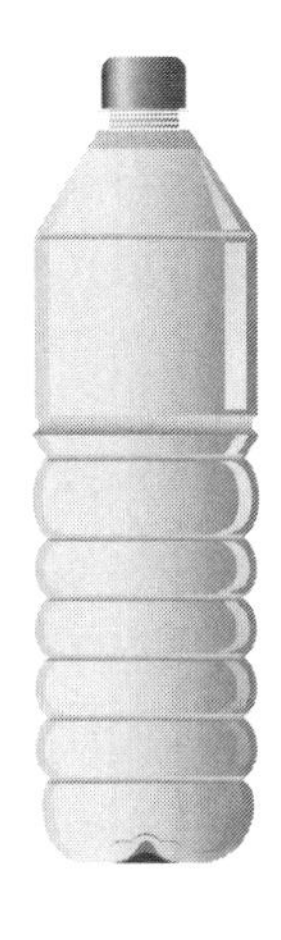

In the US, 30 billion are consumed

80% do not get recycled

Wastes more than $1 billion in plastic

2. Read aloud *One Plastic Bag*. This is a children's book that is below grade level, but it is an inspiring story of how plastic shopping bags are being repurposed.
3. Discuss the benefits of keeping materials out of landfills and reusing them or repurposing them into other products.
4. Challenge students to design a product that could be made out of plastic shopping bags, water bottles, or K-cups. Students can use one material or any combination of these materials. Allow students to work independently, in pairs, or in small groups as they respond to the design challenge. Provide general supplies such as scissors, tape, and glue so students can construct, test, and modify their designs.
5. Finish the unit by asking students to reflect on what they have learned and answer the question, *How can we reduce the volume of landfill waste?* Students may record their answers as a bulleted list or in paragraph form.

*Assess this phase:* This phase uses both formative and summative assessment. As students work on the design challenge, ask them if their design is practical, who might be interested in the product, and how many of the recyclable items they are working with would be kept out of the landfill if they were to make 10, 100, or 1,000 of their designs. Use the final response to the question, *How can we reduce the volume of landfill waste?*, as a summative assessment.

## REFERENCES

Clean Air Council. 2016. Waste and recycling facts. *http://2software.xyz/waste-and-recycling-facts-clean-air-council.html.*

Environmental Alternatives. 2016. Landfill mining. *www.enviroalternatives.com/landfill.html.*

The Math and Science Partnership Network. 2016. MiTEP list of common geoscience misconceptions organized by the Earth science literacy principles. *http://hub.mspnet.org/media/data/MiTEP_List_of_Common_Geoscience_Misconceptions.pdf?media_000000007297.pdf.*

National Geographic. 2016. Human footprint. *http://channel.nationalgeographic.com/channel/human-footprint/trash-talk2.html.*

National Governors Association Center for Best Practices and Council of Chief State School Officers (NGAC and CCSSO). 2010. *Common core state standards.* Washington, DC: NGAC and CCSSO.

NGSS Lead States. 2013. *Next Generation Science Standards: For states, by states.* Washington, DC: National Academies Press. *www.nextgenscience.org/next-generation-science-standards.*

U.S. Environmental Protection Agency (EPA). 1997. Landfill reclamation. *www.epa.gov/sites/production/files/2016-03/documents/land-rcl.pdf.*

U.S. Environmental Protection Agency (EPA). 2015. Reducing and reusing basics. *www.epa.gov/recycle/reducing-and-reusing-basics.*

U.S. Environmental Protection Agency (EPA). 2016. Advancing sustainable materials management: Facts and figures. *www.epa.gov/smm/advancing-sustainable-materials-management-facts-and-figures.*

Vijayaraghavan, A. 2011. Belgian company leads the way in landfill mining. TriplePundit. *www.triplepundit.com/2011/09/belgian-company-leads-landfill-mining.*

Name________________________________ Date____________________

# Turning Waste Into Good Business and Good Jobs

What is the message of Omar's story?

Supporting evidence

Supporting evidence

Supporting evidence

Name________________________________ Date________________

# One Year of Solid Waste

Cut out the cards below and provide one set to each group of students.

| | |
|---|---|
| Amount of trash generated by Americans in 2013 | 254 million tons of trash |
| Pounds of trash disposal per person | 4.6 pounds |
| Percentage of trash from residences | 65% |
| Percentage of trash buried in landfills | 55% |
| Percentage of trash from schools and commercial locations | 35% |
| Percentage of trash recycled | 33% |
| Percentage of trash incinerated | 12.5% |
| Number of solid-waste industry employees | 368,000 |
| Number of vehicles used to move trash to landfills | 148,000 |
| Number of communities with curbside recycling | 8,660 |
| Number of landfills | 1,754 |
| Number of recycled-materials sorting centers | 545 |
| Number of incinerators | 87 |
| Solid-waste industry annual revenue | $47 billion |

Name_______________________________ Date____________________

# Municipal Solid Waste

Cut the categories that appear at the bottom apart. Discuss each category with your team. Rank the categories from most to least amount of waste in landfills. According to your ranking, match the category cards with the percentage of municipal waste that makes sense to you. The category you think accounts for the greatest amount of solid waste in a landfill should be placed in the 27.0% section of the pie chart, and the category with the least goes in the 3.3% section.

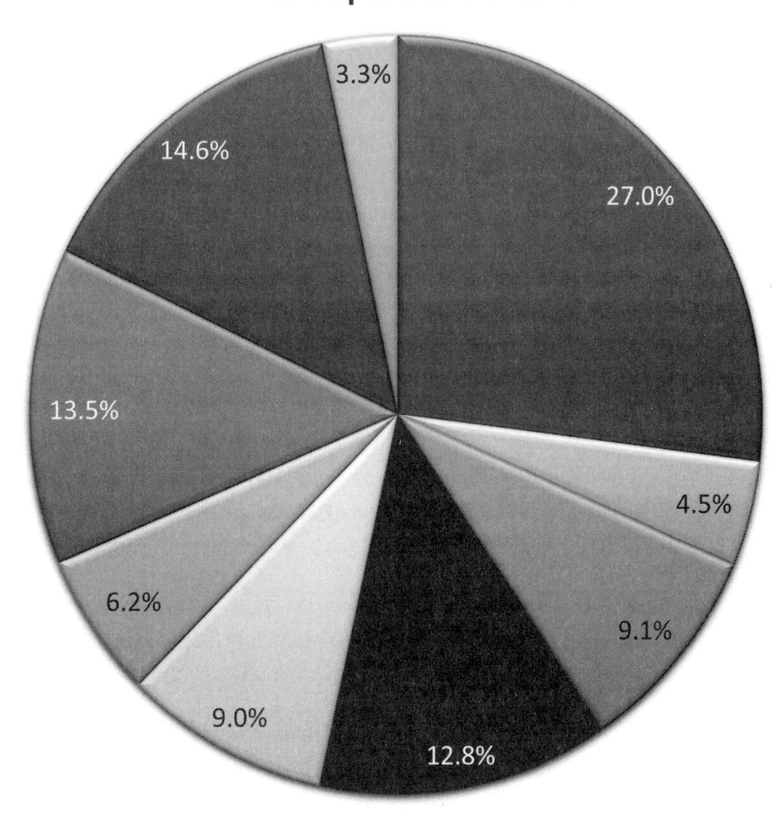


| Paper | Glass | Metals |
|---|---|---|
| Plastics | Rubber, leather, and textiles | Wood |
| Yard trimmings | Food | Other |

Name________________________ Date____________

# Landfill Mining: Brilliant Idea or Wishful Thinking?

Style Barons is an environmentally friendly company that is about to start making a new line of fashion accessories. The accessories are made of colorful plastics and precious metals. They range from belts to backpacks to bags. The accessories are made of plastic and decorated with thin strands of gold, silver, and platinum. The decorations make interesting designs that are sure to attract attention! Style Barons is going for a look and feel that seem expensive, but it wants its products to still be sold at a reasonable price.

Style Barons must decide where it is going to buy the raw materials needed for the items. Earth Materials R Us is a company that mines raw materials from Earth. Reclaiming the Past is a company that mines landfills. Style Barons needs to think about the environmental and economic impacts of the company it chooses. Of course, Style Barons also wants to make some money when it sells the accessories!

## EARTH MATERIALS R US

The upside of mining raw materials is that the new materials may be have a higher quality and have fewer impurities, and Earth Materials R Us knows where to mine. They also have an idea about how much of each raw material is available. The down side of mining raw materials is habitat destruction, hazardous waste that can pollute air and water, and dangerous working conditions.

## RECLAIMING THE PAST

The up side of mining landfills is that the unwanted waste material can be used for electricity, the landfill will last longer, and many types of recyclable materials can be recovered. Reclaiming the Past has recovered precious metals and plastics, but it cannot say for sure where it will find them in a landfill. The company also doesn't know how much of each material it will find. The downsides to mining landfills include uncovering hazardous materials, releasing landfill gases and odors, and damaging the landfill.

Style Barons has several teams of scientists researching the options. The scientists have already collected some information. The next step is to look at the information and begin forming an evidence-based argument for mining raw materials or mining landfills to recover recyclables.

Name________________________________ Date________________

## Landfill Mining: Brilliant Idea or Wishful Thinking? (*continued*)

In the graphic organizer below, list the pros and cons of each of the companies Style Barons is researching below.

| | |
|---|---|
| Earth Materials R Us | Pros |
| | Cons |
| Reclaiming the Past | Pros |
| | Cons |

## VOCABULARY

| Word | What I think it means | Connection to essay |
|---|---|---|
| | Definition | |
| Word | What I think it means | Connection to essay |
| | Definition | |
| Word | What I think it means | Connection to essay |
| | Definition | |

Questions I have

Name________________________________ Date____________________

# Based on the Evidence ...

Style Barons has several teams of scientists researching the options. The scientists have already collected some information. The next step is to look at the information and begin forming an evidence-based argument for mining raw materials or mining landfills to recover recyclables. Table 1 shows some of the environmental and economic impacts for both options. Review and discuss this information with your group before considering which company to select.

### TABLE 1. ENVIRONMENTAL AND ECONOMIC IMPACTS

| Impact Type | Regular Mining (Earth Materials R Us) | Landfill Mining (Reclaiming the Past) |
|---|---|---|
| Environmental | • Habitat loss<br>• Hazardous byproducts<br>• Soil contamination<br>• Water pollution | • Recovers recyclables<br>• Reduces landfill area<br>• Exposes hazardous materials<br>• Releases landfill gases and odors |
| Economic | • May provide a variety of jobs such as geological engineers, mining technicians, equipment operators<br>• Provides raw materials for industry | • Landfill lasts longer<br>• Produces electricity<br>• Landfills earn money selling recovered recyclables |

According to the evidence you have from "Landfill Mining: Brilliant Idea or Wishful Thinking?" and Table 1, which company do you currently think Style Barons should select?

| Evidence | Evidence |
|---|---|
| Company | |
| Evidence | Questions |

Name_______________________________________ Date_____________________

# Recovering Resources

The following memo arrived this morning:

> $^S{}_B$ Style Barons
>
> TO: Earth and Environmental Scientists Research Group
> FROM: Harper Baron, Style Barons Director of Research and Development
> SUBJECT: Sample Materials
>
> The purpose of this memo is to let you know that we are ready to test materials from Earth Materials R Us and Reclaiming the Past.
>
> Earth Materials R Us has provided us with a pure sample of one of the materials we need. Reclaiming the Past has provided us with a sample of items mined from its landfill. These items have gone through a washing process, so they are safe for handling and testing. Develop a process for sorting the materials from the landfill. Sorting is the only way we can recover what we need. After the materials are recovered, perform a test to compare the two samples.
>
> Please complete the testing within the next couple of days. We are eager to start production on the new accessories line.
>
> ITEMS SHIPPED: Pure sample from Earth Materials R Us and landfill sample from Reclaiming the Past

In response to this memo, you must do the following:

1. Work with your team to design a process for separating the landfill sample to recover the needed material. The material is a waxy substance that can be easily melted and molded. The substance will be added to the plastic that will be used for Style Barons' new line of accessories. This material will make the plastic stronger.
2. Things to consider when designing the process include the
    a. properties of the materials in the mixture,
    b. available separating equipment,
    c. purity of the sample, and
    d. quantity of the sample.
3. Write your separation plan and share it with the lab manager. The separation plan must include
    a. a separation equipment list,

Name______________________________ Date__________________

## Recovering Resources (*continued*)

b. a step-by-step procedure for separating the materials,

c. a data collection strategy to record the amount of each material recovered, and

d. an explanation of how recovered materials that are not needed will be used.

4. After obtaining the lab manager's approval, proceed with the separation of the materials. You need 30 g of the material for testing.
5. Share the results of your separation with the lab manager.
6. Obtain the sample production and testing protocol from the lab manager. For the best results, follow the protocol exactly as directed.
7. Prepare a report for Harper Baron that explains the procedure and the results.

Name________________________ Date______________

# Strength Test Data Sheet

Harper Baron is eagerly awaiting the test results. Conduct the tests as directed by the lab manager. Record all data below.

| Criteria | Pure Sample From Earth Materials R Us | Reclaimed Sample From Reclaiming the Past |
|---|---|---|
| Mass of sample | | |
| Test mass held | | |
| Ratio of test mass held to mass of sample (divide the mass held by the mass of the sample) | | |

How do the ratios of the sample and the reclaimed sample compare?

Based on this test only, which sample should Style Barons add to the plastic to make it stronger?

Based on everything you have learned about the sample from testing and about the environmental and economic effects of traditional and landfill mining, what other factors should Style Barons consider?

Name_______________________ Date_______________

# Recommendation to Style Barons

1. Working with your group, review the evidence you have gathered. Record the evidence in the table below.
2. Discuss the pros and cons of each company's mining processes and the quality of the materials that they can provide.
3. Talk about which company Style Barons should use. Write down notes during the discussion. This information could be helpful to you later.
4. Use the evidence you have gathered and meaningful information from the group discussion to make a recommendation to Style Barons.

| Considerations | Earth Materials R Us | Reclaiming the Past |
|---|---|---|
| Environmental impact | Pros | Pros |
| | Cons | Cons |
| Economic impact | Pros | Pros |
| | Cons | Cons |
| Quality of materials | Pros | Pros |
| | Cons | Cons |

Name________________________ Date________________

# Individual Recommendation to Style Barons

Using all of the evidence you have gathered and discussed with your group, make a claim about which company Style Barons should use. Present your evidence, and share your reasoning.

Claim: Which company do you recommend?

Evidence

Evidence

Evidence

Reasoning (connect the evidence to the claim)

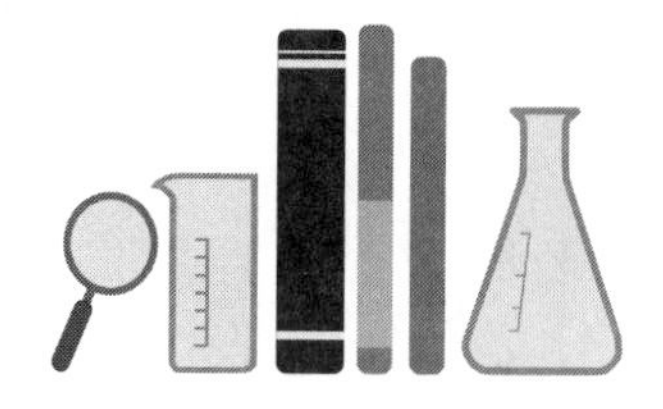

# Chapter 13
# Sunlight and the Seasons

## OVERVIEW

This unit develops an understanding of the cause of Earth's seasons. Students first make a connection between the seasons and the various amounts of sunlight associated with seasons by collecting, graphing, and analyzing sunlight data for locations in both hemispheres and at the equator. They further develop their understanding by viewing a simulation that shows Earth's axis, the angle of incidence of sunlight, and the date. This unit uses hands-on investigation and nonfiction texts to answer the questions, *How does the amount of sunlight relate to the seasons?* and *Why does Earth have seasons?*

This unit assumes that students know and can demonstrate that Earth revolves around the Sun once every 365.25 days. It also assumes that students know that the Northern Hemisphere and the Southern Hemisphere have opposite seasons. Students should be able to create line graphs to fully participate in this unit.

## OBJECTIVES

- Develop and use a model of the Sun–Earth system to explain
  - » the relationship between the amount of sunlight and the season,
  - » the relationship between the angle of incidence of sunlight and the season, and
  - » the reason Earth has seasons.
- Synthesize information from investigations, animations, and text.
- Construct a scientific argument.

## STANDARDS ALIGNMENT

### *Next Generation Science Standards* (NGSS Lead States 2013)

#### ESS1.B: EARTH AND THE SOLAR SYSTEM

- Earth's spin axis is fixed in direction over the short term but tilted relative to its orbit around the Sun. The seasons are a result of that tilt and are caused by the differential intensity of sunlight on different areas of Earth across the year.

### Common Core State Standard, English Language Arts (NGAC and CCSSO 2010)

GRADES 6–12 LITERACY IN HISTORY/SOCIAL STUDIES, SCIENCE, AND TECHNICAL SUBJECTS

- CCSS.ELA-LITERACY.RST.6-8.1: Cite specific textual evidence to support analysis of science and technical texts.
- CCSS.ELA-LITERACY.RST.6-8.2: Determine the central ideas or conclusions of a text; provide an accurate summary of the text distinct from prior knowledge or opinions.
- CCSS.ELA-LITERACY.RST.6-8.3: Follow precisely a multistep procedure when carrying out experiments, taking measurements, or performing technical tasks.
- CCSS.ELA-LITERACY.RST.6-8.7: Integrate quantitative or technical information expressed in words in a text with a version of that information expressed visually (e.g., in a flowchart, diagram, model, graph, or table).
- CCSS.ELA-LITERACY.RST.6-8.9: Compare and contrast the information gained from experiments, simulations, video, or multimedia sources with that gained from reading a text on the same topic.
- CCSS.ELA-LITERACY.WHST.6-8.1: Write arguments focused on discipline-specific content.
- CCSS.ELA-LITERACY.WHST.6-8.9: Draw evidence from informational texts to support analysis, reflection, and research.

## TIME FRAME

- Eight to nine 45-minute class periods

## SCIENTIFIC BACKGROUND INFORMATION

Earth experiences seasonal changes in temperature, weather, and hours of daylight. These changes, referred to as the *seasons,* are a result of the planet's axial tilt of approximately 23.4° relative to its orbit. Earth revolves around the Sun in a slight elliptical path once every 365.25 days. As it travels along this path, the axis points toward and away from the Sun at different times. In the Northern Hemisphere, the axis points toward the Sun the most in June. The Sun is visible for more hours of the day, and its light shines more directly and intensely, which means that more sunlight is absorbed. The greater the amount of solar energy absorbed, the more heat generated; the more heat generated, the higher the temperature. This relationship is seen in the summer in the Northern Hemisphere.

In December, Earth is in a different position with respect to the Sun, and the axis points away. The Sun appears lower in the sky, and its light is more spread out. The less direct light means that less solar energy is absorbed, and temperatures are lower. This relationship is seen in the winter in the Northern Hemisphere.

In the Southern Hemisphere, the seasons are reversed. In June, the Southern Hemisphere receives less direct sunlight and fewer hours of sunlight, which leads to lower temperatures. Thus, it is winter. In December, the Southern Hemisphere receives more direct sunlight, which leads to higher temperatures. Thus, it is summer.

Four calendar dates mark significant occurrences in seasonal change. In the Northern Hemisphere, June 21 is known as the *summer solstice*. The summer solstice marks the maximum duration and intensity of sunlight for the Northern Hemisphere. It is considered the official start of summer. December 21 is the Northern Hemisphere's winter solstice. The winter solstice marks the minimum duration and intensity of sunlight. It is the beginning of winter. September 22 (or 23) and March 21 (or 22) are the autumnal and vernal equinoxes, respectively. On those days, the location of Earth in its orbit means that the axis is oriented at a 90° angle with respect to the Sun. Nearly every location on Earth receives approximately 12 hours of sunlight. In fact, the word *equinox* means "equal night" in Latin. The exceptions are the North and South Poles, for which the equinoxes are the sunrise and sunset after the polar night and day.

The effects of axial tilt are more pronounced at higher latitudes. The polar regions are famous for having prolonged periods of daylight in the summer and prolonged periods of darkness in the winter. In contrast, locations near the equator have nearly equal day and night all year round, as well as fairly constant temperatures. This is because the equator experiences little to no variation in the intensity and angle of sunlight.

## MISCONCEPTIONS

A widely held and well-documented misconception is that Earth is closer to the Sun in summer than it is in winter (Atwood and Atwood 1996; Dove 1998; Philips 1991; Sadler 1998). This erroneous idea is prevalent among students and adults alike. In a famous (or infamous) video, *A Private Universe*, graduates at Harvard University's commencement failed to correctly explain the cause of seasonal change (Schneps and Sadler 1989). Other common misconceptions are presented in Table 13.1.

### TABLE 13.1. COMMON MISCONCEPTIONS ABOUT SEASONS

| Common Misconception | Scientifically Accurate Concept |
| --- | --- |
| • Earth is closer to the Sun in summer, so it is warmer. Earth is farther away from the Sun in winter, so it is colder.<br>• Earth leans toward the Sun in the summer and away from the Sun in the winter. | Earth experiences seasons because of its axial tilt relative to its orbit around the Sun. This tilt leads to changes in the intensity of the sunlight received throughout the year. |
| The side of Earth not facing the Sun experiences winter. | With this misconception, students are confusing the daily rotation of Earth with the yearly revolution of Earth around the Sun. |

*Source:* Galili and Lavrik 1998; Sadler 1987; Salierno, Edelson, and Sherin 2005.

## MATERIALS

- Computers with Adobe Flash Player and internet access
- Interactive white board (optional)
- Graph paper (optional)
- Chart paper
- Lamp
- Styrofoam balls (1 per small group of students)

## SAFETY CONSIDERATION

- Remind students that the lamp bulbs may get very hot and can burn skin.

## SUPPORTING DOCUMENTS

- "*Arctic Lights, Arctic Nights* Data Table"
- "Seasons and Sunlight" activity sheet
- "Making Connections I" graphic organizer
- "Making Connections II" graphic organizer
- "Writing an Argument About Seasons" planning sheet

## NONFICTION TEXTS

*Arctic Lights, Arctic Nights* by Debbie S. Miller (London: Walker, 2007); Flesch-Kincaid reading level 4.9, published Lexile level 890L.

Explore the monthly changes in daylight, temperature, and environment just south of the Arctic Circle. This book is a great introduction to the solstices and equinoxes that occur throughout the year. Beautiful illustrations accompany the descriptive text.

*Earth* by Stuart Clark (Portsmouth, NH: Heinemann, 2008); Flesch-Kincaid reading level 5.4, published Lexile level 870L.

This comprehensive book provides information about seasons and Earth's atmosphere. It also examines the structure and formation of Earth and the history of life on Earth. The text is accessible and straightforward.

QR CODE 1

"Seasons" by CK–12 Foundation, *http://bit.ly/1VhpLd7;* Flesch-Kincaid reading level 5.5. (QR code 1)

This brief article explains the cause of Earth's seasons. Diagrams and images of where sunlight reaches Earth on the solstices and equinoxes complement the straightforward text.

QR CODE 2

"What Causes the Seasons?" by NASA SpacePlace, *http://go.nasa.gov/1KnCqaQ;* Flesch-Kincaid Reading Level 6.5. (QR code 2)

This brief article explains the cause of Earth's seasons. It also addresses the common misconception that Earth is closer to the Sun in the summer and further away in the winter. Diagrams show Earth's slightly lopsided orbit and its position with respect to the Sun in December, March, June, and September.

## LEARNING-CYCLE INQUIRY

### Engage

In this phase, students are presented with an unusual video of the Sun's path across the sky. Discussion about this video elicits their prior understanding of the Sun and seasons.

1. Engage students in conversation about the Sun and what they have noticed about it recently. Have the days been getting shorter or longer? Ask students to share their ideas about why the amount of sunlight varies throughout the year. Do they notice a connection between the season and the amount of sunlight? What about between the temperature and the season?

QR CODE 3

2. Show students the video "Sun Returns to Igoolik" *(http://safeshare.tv/v/WPM3EjbDuSk;* QR code 3*)*. This video shows an early spring day in Igoolik, a small village in the Canadian Arctic. Locate the village on a map before showing students the video. Explain to them that the video was taken on an early spring day. After they watch the video, ask students to discuss their observations and ideas about the video. Why do they think the Sun is moving this way? What do you think the video title means by "Sun returns"?

3. Explain to students that, in this inquiry, they will investigate the connection between sunlight and seasons. Introduce the following questions: *How does the amount of sunlight relate to the seasons? Why does Earth have seasons?*

***Assess this phase:*** Only formative assessment is needed at this time. Monitor student participation, and pay attention to students' ideas about sunlight and seasons.

## Explore

In this phase, students collect several types of data (including hours of daylight and average temperature) for three locations using a nonfiction text and websites. They analyze this data and use it to generate evidence-based claims. (*Note:* This activity is adapted, with permission, from "Teaching Through Trade Books: Seasons by the Sun" by Meri-Lyn Stark, which appeared in the July 2005 issue of *Science and Children.*)

### PART I: SEASONAL CHANGE IN FAIRBANKS, ALASKA

1. Introduce the book *Arctic Lights, Arctic Nights* to students. Locate Fairbanks, Alaska, on a map and explain that this book describes changes that occur throughout the year in an unspecified location near Fairbanks. Tell students that they will be listening to the book and collecting data from the text.
2. Pass out the "*Arctic Lights, Arctic Nights* Data Table" handout. Explain to students that for each date given in the book, they will record the time of sunrise and sunset, the amount of daylight in hours and minutes, the high and low temperatures, and any additional information that they deem important. (*Note:* You may wish to support students by recording this data on an interactive white board, overhead, or standard white board during the reading of the text.)
3. Read the text aloud, pausing after each two-page spread to allow students to record data. Draw students' attention to the qualitative descriptions of the environment and the illustrations as you go, inviting students to record information that they think is important. Additionally, have students label the data for June 21 as *summer solstice,* September 21 as *autumnal equinox,* December 21 as *winter solstice,* and March 21 as *vernal equinox.* (*Note:* Students may wish to discuss the text or ask questions as you read. Allow them to share, but do not give them explanations for the wide range of sunlight or environmental conditions. Instead, explain to them that they will be working to answer such questions throughout the inquiry.)
4. After you finish reading the text, have students construct a line graph showing the amount of daylight by month. To be consistent with the text, have students begin their graphs with June 21 and end them with May 21. Both the data table and the graph should be pasted into students' science notebooks, if applicable. (*Note:* If your students are not proficient in graphing, you may choose to conduct a mini-lesson on

creating graphs before asking them to construct the graph. Alternatively, you may lead the class through the construction of the graph, modeling work on the board while students create their own copies.)

5. Ask students to review their data tables and graphs and to think about the patterns they observe. Ask them what claims they can make and what evidence supports these claims. Allow students to think-pair-share.
6. Have students write claims and evidence in a T-chart in their science notebooks or on notebook paper. Encourage students to make as many claims as possible, as long as they can be supported by evidence from the Explore phase.
7. Conduct a class discussion in which you invite students to share their claims and evidence. Record these claims in a computer file, on chart paper, or in another way for later use.
8. Call students' attention to the labels you added to June 21, September 21, December 21, and March 21. Ask the following questions: What is a *solstice?* What is an *equinox?* Invite students to discuss the terms in pairs or small groups and to generate working definitions for them. Share ideas with the whole class and post collaboratively generated definitions in the classroom.

### PART II: SEASONAL CHANGE IN YOUR TOWN

QR CODE 4

1. Explain to students that they will collect the same data for a new location: your city or town. At this time, students should construct their own data table, using the table from *Arctic Lights, Arctic Nights* as a model. Students will use websites to collect sunlight *(http://bit.ly/1hM8aCm)* and temperature *(http://bit.ly/1jN60G0)* data. (See QR codes 4 and 5.)

QR CODE 5

2. Students will create another line graph showing the amount of sunlight per month for your location. Have students begin their graphs with June 21 and use the same intervals for the *y*-axis. This consistency will allow students to make direct comparisons between the two graphs.
3. Ask students to review their data tables and graphs for your location. What patterns do they observe?
4. Students will record claims and evidence in a T-chart like the one they created for the Arctic.
5. Ask students to compare their graphs and data tables for the Arctic with those for their location. What similarities and differences do they observe? Have them add claims and evidence to their T-charts.

6. Have a class discussion about students' work. Record their claims and evidence in a place that can be accessed later in the inquiry.

### PART III: SEASONAL CHANGE AROUND THE WORLD

1. Explain to students that they will collect the same data for a third location. Assign students to research one of several locations: Barrow, Alaska; Bogotá, Colombia; McMurdo Station, Antarctica; and Santiago, Chile. Students should create another data table as they did in Part II. They will continue to use the same websites to collect sunlight and temperature data for their third location. Students will create another line graph showing the amount of sunlight per month for the location they researched. Have students begin their graphs with June 21 and use the same intervals for the $y$-axis. This consistency will allow students to make direct comparisons between the two graphs.
2. Ask students to review their data tables and graphs. What patterns do they observe?
3. Students will record claims and evidence in a T-chart like the one they created in Parts I and II.
4. Ask students to compare their graphs and data tables for the Arctic Circle and their locations. What similarities and differences do they observe? Have students add their claims and evidence to their T-charts.
5. Have a class discussion about students' work. Record their claims and evidence in a place that can be accessed later in the inquiry.

***Assess this phase:*** As in the previous phases, only formative assessment is needed. During Part I, monitor student engagement with the text, and check data tables to ensure that students have recorded data correctly. Some students (such as those with attention or learning difficulties) may benefit from having a partially completed table in which dates and some data are pre-recorded for them. If students struggle to complete the initial line graph, take the time to teach and provide support for this skill, because students will complete several more line graphs during the inquiry. Students' written work (their claims and evidence) provides insight into their developing understanding of the content, as well as their ability to analyze data. Collecting and reading this work is essential at this point in the inquiry because the subsequent phases build on this first portion. If students struggle to generate evidence-based claims, consider providing an example or conducting this phase as a whole-class activity. During Parts II and III, continue to monitor student work.

## Explain

In this phase, students visualize the data collected in the Explore phase graphically and summarize their understanding of seasonal change in the Northern Hemisphere, in the Southern Hemisphere, and at the equator.

1. Display a large world map that all students can see. A blank map projected on an interactive white board is a great choice. You can also use a poster or drawing on butcher paper. Mark each location studied on the map, and title the map "Hours of Sunlight on June 21."
2. Begin with the northernmost location (Barrow, Alaska). Ask students to share the number of hours of sunlight the area receives on June 21, and add that number to the map.
3. Repeat step 2 for all locations, moving from north to south.
4. Ask students to look for patterns in the data when displayed on a map. Have them add claims and evidence to a T-chart.
5. Discuss students' work, and post their claims and evidence in a place that can be easily accessed throughout the inquiry.
6. Repeat steps 1–5 for the December 21, March 21, and September 21 data.
7. Discuss what students have learned about seasonal changes so far. Guiding questions such as the following may be helpful in eliciting students' ideas and extending their thinking:
   - On June 21, what season is it where we live? How do you know? How do the data you collected support your answer?
   - On June 21, what season is it at McMurdo Station, Antarctica? How do you know? How do the data you collected support your answer?
   - How do the seasons of the Northern Hemisphere compare with those of the Southern Hemisphere? How do the data support your answer?
   - How do the seasons near the equator compare with those of the Northern and Southern Hemispheres? How do the data support your answer?
   - Do you see a connection between the number of hours of daylight and the average temperature of a place? Explain your thinking.
8. Have students write a paragraph summarizing their thinking about seasonal change around the world over the course of a year. Allow students to reference all materials while writing.

*Assess this phase:* Although formal writing occurs in this phase, it leads to a larger assignment later on in the unit and is thus considered a piece of formative assessment. At this point in the unit, students should be able to connect the number of hours of daylight to the seasons. They should also be able to explain that seasons are reversed in the Southern Hemisphere, and that locations near the equator do not undergo seasonal variation in the amount of sunlight and average temperature. Students may also begin to connect the number of hours of daylight and average temperature. If students are not making these connections or are unable to explain the connections in sufficient detail, consider repeating the activities in a more teacher-directed manner.

## Explore

In this phase, students use a simulation of Earth as it orbits the Sun to connect the seasons with Earth's position. Students also read nonfiction text to solidify their understanding of the connections among Earth's position, the movement of Earth around the Sun, the number of hours of daylight, and the season.

QR CODE 6

1. Preview the animation found at *http://bit.ly/1AYB0ZB* (QR code 6) with students. Discuss the types of information displayed in the three sections: (1) the position of Earth relative to the Sun, (2) an observer's latitude, and (3) the angle of incidence for incoming sunlight. Move the observer to your approximate latitude and play the simulation through at least once. You may also wish to discuss inaccuracies in the animation with students. For example, the animation presents a simplified view of the Sun–Earth relationship without the Moon or other planets, and the animation is not drawn to scale. You may also need to introduce your students to the concept of Earth's axis before proceeding.
2. Distribute the student sheet "Seasons and Sunlight," or ask students to construct a similar page in their science notebooks. Students will work independently or in small groups to explore the animation, depending on the technology available. The procedure found on the student sheet is as follows:
    a. Students will change the location of the observer by dragging the red line to the North Pole.
    b. Students will play the animation, pausing at March 21.
    c. Students will sketch the position of Earth and the Sun and the angle at which sunlight hits Earth. They will note what season it is at that location.
    d. Students will repeat steps b and c for June 21, September 21, and December 21.
    e. Students will change the location of the observer and repeat steps a–d for three more locations: (1) your location's latitude, (2) the equator, and (3) the South Pole.
    f. Students will generate claims and evidence about the patterns they observe in the data and record them in a T-chart.
3. Discuss students' work, and post their claims and evidence in a place that can be easily accessed throughout the inquiry.

4. Have students write about the connections they make between this data and the daylight and temperature data from earlier phases.
5. Explain to students that they will use nonfiction texts to help them further develop their understanding. Read aloud pages 8–9 of *The Universe: Earth.* Please note that in the diagram on page 8, Earth's orbit is depicted as extremely elliptical, when it is actually much more circular. We recommend using this book as a read-aloud and not showing the diagram to students, which can inadvertently strengthen the misconception that Earth is closer to the Sun in summer and farther away in winter. As you read, use a copy of the "Making Connections I" graphic organizer to model how to take notes from the text in the left-hand column of the organizer.
6. Ask students to share information from the text that connects to what they discovered in the Explore phases. Record this information in the right-hand column of the "Making Connections I" graphic organizer.
7. Have students read the article "What Causes the Seasons?" Distribute copies of the "Making Connections I" graphic organizer for students to use as they read. Walk around the room as students work, and provide support as needed.
8. Ask students to share the connections they made between their investigations and the article.
9. Have students read the article "Seasons." Instruct students to use the "Making Connections II" graphic organizer to record new information and to make connections between this article, their previous reading, and the investigations.
10. Ask students to share the new information they learned from the article as well as connections they made between their investigations and the articles. Collect both "Making Connections" graphic organizers as a means of formative assessment.

***Assess this phase:*** Formative assessment is particularly critical in this phase, because students are expected to demonstrate their understanding in the next phase of the unit. Students should be able to articulate that the position of Earth and its tilt is directly connected to the angle of incidence of sunlight, the number of hours of daylight, and the average temperature. If students are unable to make these connections, further exploration may be needed. Suggestions for further exploration include kinesthetic activities such as walking a Styrofoam ball (Earth) around a lamp (the Sun) and pausing to note its position at the solstices and equinoxes and creating drawings that depict Earth's position at each of these four dates. Students' contributions to class discussions and the worksheets they complete after viewing the videos also provide a means of formative assessment.

## Explain

In this phase, students demonstrate their understanding by developing a scientific argument. They will write a scientific argument to answer the question, *Why does Earth have seasons?* Students will follow the claims, evidence, and reasoning (C-E-R) framework (McNeil and Krajcik 2011) to write this.

The C-E-R framework is a strategy for constructing a scientific argument. It includes three components:

- Claim: a statement or conclusion that answers the question
- Evidence: appropriate and sufficient scientific data to support the claim
- Reasoning: a justification that shows why the data count as evidence and includes scientific principles to connect the evidence and claim

You can introduce students to writing with the C-E-R framework in a variety of ways. Students will have already had experience with claims and evidence from the Explore phase of this unit. They will most likely need support in formulating the reasoning piece of the framework and in putting the parts together to form a cohesive argument. In this activity, we use a template (adapted with permission from Buell 2012) to assist students with this process.

1. Distribute the "Writing an Argument About Seasons" student sheets.
2. Read over the sheets with students. Provide an explanation for each section of the template, and allow students to work independently. Support individual students as needed. If students have difficulty with the reasoning section, remind them to use the scientific principles and terminology from the article to connect the evidence to their claim. Students should also have access to their work from the Explore phase while constructing their arguments. You may choose to have students rewrite their argument as a single paragraph, or you may accept the completed template as the finished product. This choice will ultimately depend on your time constraints and your students' level of comfort with argumentation.

***Assess this phase:*** Completed worksheets associated with the videos serve as a final formative assessment. The completed argument (whether in template or paragraph form) is one piece of summative assessment for the unit. It should be assessed for content knowledge; that is, students should demonstrate the correct understanding that the tilt of Earth's axis affects the amount and intensity of sunlight received at a given location and thus the temperature of that location. If students are new to argumentation, we do not recommend that you assess them on their ability to write a scientific argument. However, if students are sufficiently familiar with argumentation, you may choose to evaluate their work on this criteria.

### Expand

In this phase, students will further demonstrate their understanding of seasons by considering an alternate scenario—if Earth were not tilted on its axis.

1. Ask students, "What would happen to Earth's seasons if Earth were not tilted on its axis?" Allow students to think-pair-share for a few moments.
2. Invite small groups of students to explore this question by using a physical model to simulate Earth's rotation around the Sun. Distribute Styrofoam balls to small groups, and set up a lamp on a desk in an area around which students can move.
3. Students should first model Earth's movement around the Sun with its current axial tilt. As they move around the lamp, students should discuss how the tilt of Earth connects to seasonal changes.
4. Have students hold the Styrofoam ball straight up and down to model Earth without an axial tilt. Have groups move around the lamp, observing carefully. Small groups should discuss their observations.
5. Students will write a paragraph answering the question from step 1. Students should write independently and also have access to all materials, books, and websites during this phase. Some students may need to repeat the kinesthetic activity individually or with teacher assistance.

*Assess this phase:* Student conversation in the whole-class and small-group discussions provides another opportunity for formative assessment. The written paragraph serves as a second piece of summative assessment for the unit. Students who are secure in their understanding of seasons should be able to articulate that Earth would not experience seasonal change without an axial tilt. Locations across Earth would still experience different temperatures and amounts of daylight, but these would be constant over the course of the year for a given location.

## REFERENCES

Atwood, R. K., and V. A. Atwood. 1996. Preservice elementary teachers' conceptions of the causes of seasons. *Journal of Research in Science Teaching* 33 (5): 553–563.

Buell, J. 2012. Claim evidence reasoning. Always Formative. *http://alwaysformative.blogspot.com/2012/04/claim-evidence-reasoning.html.*

Dove, J. 1998. Alternative conceptions about the weather. *School Science Review* 79 (289): 65–69.

Galili, I., and V. Lavrik. 1998. Flux concept in learning about light: A critique of the present situation. *International Journal of Science Education* 82 (5): 591–613.

McNeil, K. L. and J. S. Krajcik. 2011. *Supporting grade 5–8 students in constructing explanations in science: The Claim, Evidence, and Reasoning Framework for talk and writing.* Boston, MA: Pearson.

National Governors Association Center for Best Practices and Council of Chief State School Officers (NGAC and CCSSO). 2010. *Common core state standards.* Washington, DC: NGAC and CCSSO.

NGSS Lead States. 2013. *Next Generation Science Standards: For states, by states.* Washington, DC: National Academies Press. *www.nextgenscience.org/next-generation-science-standards.*

Philips, W. 1991. Earth science misconceptions. *Science Teacher* 58 (2): 21–23.

Sadler, P. 1987. Misconceptions in astronomy. In *Proceedings of the second international seminar misconceptions and educational strategies in science and mathematics, volume III,* ed. J. Novak, 422–425. Ithaca, NY: Cornell University.

Sadler, P. 1998. Psychometric models of student conceptions in science: Reconciling qualitative studies and distractor-driven assessment items. *Journal of Research in Science Teaching* 35 (3): 265–296.

Salierno, C., D. Edelson, and B. Sherin. 2005. The development of student conceptions of the Earth–Sun relationship in an inquiry-based curriculum. *Journal of Geoscience Education* 53 (4): 422–431.

Schneps, M., and P. Sadler. 1989. *A private universe* [video]. Santa Monica, CA: Pyramid Film and Video.

Stark, M. 2005. Seasons by the Sun. *Science and Children* 42 (8): 14–16.

Name________________________ Date________________

# Arctic Lights, Arctic Nights Data Table

In the table below, record the time of sunrise and sunset, the amount of daylight in hours and minutes, the high and low temperatures, and any additional information.

| Date | Sunrise | Sunset | Amount of Daylight | High Temperature | Low Temperature | Other Information |
|---|---|---|---|---|---|---|
| | | | | | | |
| | | | | | | |
| | | | | | | |
| | | | | | | |
| | | | | | | |

Name____________________ Date________________

*Arctic Lights, Arctic Nights Data* Table (*continued*)

| Date | Sunrise | Sunset | Amount of Daylight | High Temperature | Low Temperature | Other Information |
|---|---|---|---|---|---|---|
| | | | | | | |
| | | | | | | |
| | | | | | | |
| | | | | | | |
| | | | | | | |
| | | | | | | |
| | | | | | | |

Name______________________________ Date______________

# Seasons and Sunlight

Use the animation found at *http://bit.ly/1AYB0ZB* (see QR code) to collect data about Earth, the Sun, and the seasons.

QR CODE

1. Find the box in the upper right-hand corner of the screen. Change the location of the observer by dragging the red line to the North Pole.
2. Play the animation (beginning with January), and observe the movement of the Sun. Pause the animation at March 21 by clicking on the "Stop Animation" button.
3. In the table below, sketch the position of Earth and the Sun and the angle at which sunlight hits Earth. Note what season it is at that location.
4. Repeat steps 2 and 3 for June 21, September 21, and December 21.

### LOCATION: NORTH POLE

| Date | March 21 | June 21 | September 21 | December 21 |
|---|---|---|---|---|
| Sketch of Earth and the Sun | | | | |
| Sketch of angle of sunlight | | | | |
| Season | | | | |

Name_______________________________________________ Date____________________________

## Seasons and Sunlight *(continued)*

5. Change the location of the observer by dragging the red line to your location.
6. Play the animation (beginning with January), and observe the movement of the Sun. Pause the animation at March 21.
7. In the table below, sketch the position of Earth and the Sun and the angle at which sunlight hits Earth. Note what season it is at that location.

**LOCATION: ________________________________ (WRITE YOUR LOCATION)**

| Date | March 21 | June 21 | September 21 | December 21 |
|---|---|---|---|---|
| Sketch of Earth and the Sun | | | | |
| Sketch of angle of sunlight | | | | |
| Season | | | | |

8. Repeat steps 5 and 6 for June 21, September 21, and December 21.
9. Change the location of the observer by dragging the red line to the equator.
10. Play the animation (beginning with January) and observe the movement of the Sun. Pause the animation at March 21.
11. In the table below, sketch the position of Earth and the Sun and the angle at which sunlight hits Earth. Note what season it is at that location.
12. Repeat steps 9 and 10 for June 21, September 21, and December 21.

Name______________________________ Date__________________

## Seasons and Sunlight (*continued*)

### LOCATION: THE EQUATOR

| Date | March 21 | June 21 | September 21 | December 21 |
|---|---|---|---|---|
| Sketch of Earth and the Sun | | | | |
| Sketch of angle of sunlight | | | | |
| Season | | | | |

13. Change the location of the observer by dragging the red line to the South Pole.
14. Play the animation (beginning with January), and observe the movement of the Sun. Pause the animation at March 21.
15. In the space below, sketch the position of Earth and the Sun and the angle at which sunlight hits Earth. Note what season it is at that location.
16. Repeat steps 13 and 14 for June 21, September 21, and December 21.

Name________________________ Date________________

**Seasons and Sunlight** (*continued*)

### LOCATION: THE SOUTH POLE

| Date | March 21 | June 21 | September 21 | December 21 |
|---|---|---|---|---|
| Sketch of Earth and the Sun | | | | |
| Sketch of angle of sunlight | | | | |
| Season | | | | |

17. What claim(s) can you make about the position of Earth, the angle of sunlight, and the season? What evidence supports your claim(s)?

| Claim | Evidence |
|---|---|
| | |

Name______________________________ Date__________________

# Making Connections I

| What I learned from the article | How the article connects to my investigations |
|---|---|
| | |

Name________________________________ Date________________

# Making Connections II

| New information from the article | How the article connects to other articles and my investigation |
|---|---|
| | |

Name______________________________ Date____________________

# Writing an Argument About Seasons

Write a scientific argument that answers the question, *Why does Earth have seasons?*

## CLAIM

(Write a sentence that states why Earth has seasons.)

## EVIDENCE

(Provide data that support your claim about why Earth has seasons.)

## REASONING

(Write a statement that connects your evidence to your claim about why Earth has seasons.)

**Chapter 14**

# Getting to Know Geologic Time

## OVERVIEW

This unit develops an understanding of geologic time and how it is organized and described by the geologic time scale. Students learn how fossils and rock strata are used to organize Earth's history. This unit uses hands-on investigation and nonfiction texts to answer the question, *How can we use geologic evidence to represent Earth's history?*

This unit assumes that students know and can demonstrate an understanding of a simple timeline. Students should have mastered this skill in the elementary grades, but if students seem to lack experience with simple timeline, spend some time creating a timeline of events that they can relate to before beginning this unit.

## OBJECTIVES

- Explain that Earth's history is broken up into smaller time periods.
- Explain that rocks and fossils provide evidence for these time periods.
- Synthesize information gathered through investigations and text.
- Construct a scientific explanation based on evidence.

## STANDARDS ALIGNMENT

### *Next Generation Science Standards* (NGSS Lead States 2013)

#### ESS1.C: THE HISTORY OF PLANET EARTH

- The geologic time scale interpreted from rock strata provides a way to organize Earth's history. Analyses of rock strata and the fossil record provide only relative dates, not an absolute scale. (MS-ESS1-4)

### *Common Core State Standards, English Language Arts* (NGAC and CCSSO 2010)

#### GRADES 6–12 LITERACY IN HISTORY/SOCIAL STUDIES, SCIENCE, AND TECHNICAL SUBJECTS

- CCSS.ELA-LITERACY.RST.6-8.7: Integrate quantitative or technical information expressed in words in a text with a version of that information expressed visually (e.g., in a flowchart, diagram, model, graph, or table).
- CCSS.ELA-LITERACY.RST.6-8.9: Compare and contrast the information gained from experiments, simulations, video, or multimedia sources with that gained from reading a text on the same topic.

- CCSS.ELA-LITERACY.WHST.6-8.2: Write informative/explanatory texts, including the narration of historical events, scientific procedures/experiments, or technical processes.
- CCSS.ELA-LITERACY.WHST.6-8.4: Produce clear and coherent writing in which the development, organization, and style are appropriate to task, purpose, and audience.
- CCSS.ELA-LITERACY.WHST.6-8.9: Draw evidence from informational texts to support analysis, reflection, and research.

## TIME FRAME

- Eight to ten 45-minute class periods

## SCIENTIFIC BACKGROUND INFORMATION

Earth is 4.6 billion years old, give or take 50 million years. Although the oldest rocks have been dated to only 4.0 billion years old, geologists have dated zircon crystals (a type of mineral) from the Jack Hills in Western Australia to 4.4 billion years old (Rosen 2015). Of this time, life has existed on Earth for approximately 3.8 billion years, although recent studies suggest that living organisms may have been present for even longer.

Fossils and rocks provide the geologic evidence for understanding the history of life on Earth. The position of fossils within rock layers, or *strata,* allows scientists to determine the relative age of fossils. The law of superposition states that older rock strata are found beneath younger rock strata. Of course, scientists are now also able to determine the absolute age of rocks and fossils through radiometric dating methods (i.e., determining age by analyzing the decay of radioactive elements present in the specimen). Both relative and absolute dating have allowed scientists to piece together the history of Earth and of life on it.

Geologists organize and describe Earth's history using the geologic time scale. In this scale, the geological history of Earth is broken up into categories: eons, eras, periods, epochs, and ages. These categories are hierarchical, so eons are the largest division of time, and ages are the smallest. There is one supereon, the Precambrian, which includes the Hadean, Archean, and Proterozoic eons. The time scale is represented as a vertical timeline, with the oldest eon at the bottom and the most recent at the top. Figure 14.1 shows one representation of the geologic time scale.

It is important to note that the time scale in Figure 14.1, which is a common representation, is a variable time scale and does not visually represent the duration of each time period accurately. The Precambrian, for example, represents approximately 90% of Earth's history, but the only life forms in existence were single-celled bacteria (during the Archaen) and simple, multicellular life (during the Proterozoic). Compare the variable time scale with the constant time scale in Figure 14.2. This version shows a to-scale representation of the Precambrian and subsequent eras.

The Precambrian ended 542 million years ago. That date marks the beginning of a new eon, the Phanerozoic, when living things evolved at a rapid rate. The first period of the

**FIGURE 14.1. GEOLOGICAL TIME SCALE**

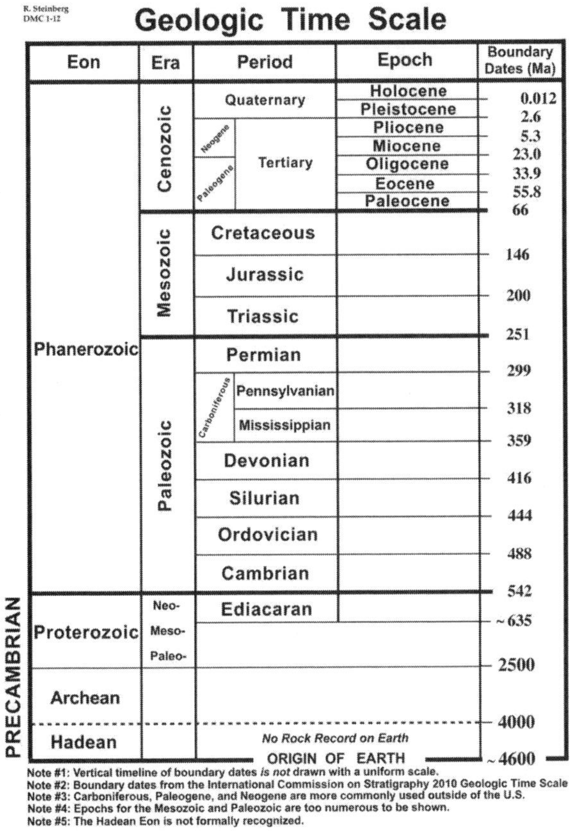

R. Steinberg
DMC 1-12

Geologic Time Scale

| Eon | Era | Period | Epoch | Boundary Dates (Ma) |
|---|---|---|---|---|
| Phanerozoic | Cenozoic | Quaternary | Holocene | 0.012 |
| | | | Pleistocene | 2.6 |
| | | Tertiary (Neogene) | Pliocene | 5.3 |
| | | | Miocene | 23.0 |
| | | (Paleogene) | Oligocene | 33.9 |
| | | | Eocene | 55.8 |
| | | | Paleocene | 66 |
| | Mesozoic | Cretaceous | | 146 |
| | | Jurassic | | 200 |
| | | Triassic | | 251 |
| | Paleozoic | Permian | | 299 |
| | | (Carboniferous) Pennsylvanian | | 318 |
| | | Mississippian | | 359 |
| | | Devonian | | 416 |
| | | Silurian | | 444 |
| | | Ordovician | | 488 |
| | | Cambrian | | 542 |
| PRECAMBRIAN: Proterozoic | Neo- | Ediacaran | | ~635 |
| | Meso- | | | |
| | Paleo- | | | 2500 |
| Archean | | | | 4000 |
| Hadean | | No Rock Record on Earth | | ~4600 |

ORIGIN OF EARTH

Note #1: Vertical timeline of boundary dates *is not* drawn with a uniform scale.
Note #2: Boundary dates from the International Commission on Stratigraphy 2010 Geologic Time Scale
Note #3: Carboniferous, Paleogene, and Neogene are more commonly used outside of the U.S.
Note #4: Epochs for the Mesozoic and Paleozoic are too numerous to be shown.
Note #5: The Hadean Eon is not formally recognized.

**FIGURE 14.2. CONSTANT TIME SCALE**

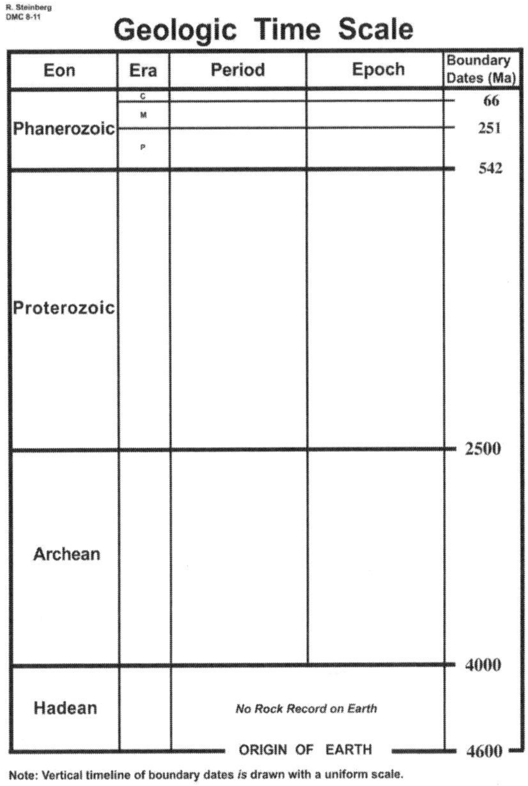

R. Steinberg
DMC 8-11

Geologic Time Scale

| Eon | Era | Period | Epoch | Boundary Dates (Ma) |
|---|---|---|---|---|
| Phanerozoic | C | | | 66 |
| | M | | | 251 |
| | P | | | 542 |
| Proterozoic | | | | 2500 |
| Archean | | | | 4000 |
| Hadean | | No Rock Record on Earth | | 4600 |

ORIGIN OF EARTH

Note: Vertical timeline of boundary dates *is* drawn with a uniform scale.

Phanerozoic, the Cambrian Period, is the time when most of the major groups of animals first appear in the fossil record. Geologists often refer to this time as the *Cambrian explosion* because of the relatively short time period, geologically speaking, over which this diversity of life appeared.

Scientists use major biological and geological events, such as the Cambrian explosion or the extinction of the dinosaurs, to divide eons, eras, periods, epochs, and ages. The evidence for these events is contained in rock strata and in the fossil record.

## MISCONCEPTIONS

Geologic time is a difficult concept for students to understand. They may find it hard to conceptualize the vast amount of time and work with the large numbers involved. In addition, students often arrive in middle school with knowledge of only dinosaurs, having never learned about other types of ancient plants and animals. Table 14.1 (p. 290) lists some common misconceptions about geologic time.

## TABLE 14.1. COMMON MISCONCEPTIONS ABOUT GEOLOGIC TIME

| Common Misconception | Scientifically Accurate Concept |
|---|---|
| Life on Earth existed as soon as Earth formed. | The Hadean eon, which spanned 4.6 billion to 4.0 billion years ago, was a time where life did not exist on Earth. The earliest evidence of living organisms is from 3.8 billion years ago, although recent studies suggest it may have originated earlier than that. Nonetheless, there was a significant period without life early in Earth's history. |
| Earth had a single continent when humans first appeared on Earth. | Students who hold this misconception may be incorrectly associating the appearance of humans with the existence of Pangaea, which was a supercontinent that existed between 300 million and 175 million years ago. The oldest human remains date to about 3 million years old. |
| Dinosaurs appeared about halfway through Earth's history. | Dinosaurs first appeared about 230 million years ago, during the Triassic period. This is much later than halfway through Earth's 4.6 billion year history. |

*Source:* Dahl, Anderson, and Libarkin 2005; Libarkin and Anderson 2005; Libarkin et al. 2005.

## NONFICTION TEXTS

*Evolution: How We and All Living Things Came to Be* by Daniel Loxton (Toronto, ON: Kids Can Press, 2010); Flesch-Kincaid reading level 7.6, published Lexile level 1060L.

Through the use of relatively short passages written in plain language, coupled with colorful illustrations, the author presents evolution in a way that is clear and scientifically accurate. Common questions and misconceptions about evolution are addressed in a thorough and respectful manner.

QR CODE 1

Online

"Geologic Time Scale" by Encyclopedia Britannica for Kids, *http://bit.ly/23Zvoj8;* Flesch-Kincaid reading level 7.5. (QR code 1)

This entry provides a full-color, simplified version of the geologic time scale that is appropriate for middle school students.

# GETTING TO KNOW GEOLOGIC TIME

# 14

"Understanding Geologic Time" by University of California Museum of Paleontology and National Science Foundation, *http://bit.ly/1gw3NhJ*; Flesch-Kincaid reading level ranges from 3.6 to 7.7. (QR code 2)

QR CODE 2

This interactive module includes text and images and allows students to work through a number of "big ideas" about timelines, stratigraphy, and geologic time.

*Wild Horse Scientists* by Kay Frydenborg (Boston, MA: Houghton Mifflin Harcourt Books for Young Readers, 2012); Flesch-Kincaid reading level 8.4, published Lexile level 1210L.

This book documents the work two scientists are doing with the wild horses of Assateague Island National Seashore. Beautiful photographs, descriptive text, and the scientists' handwritten notes offer a glimpse into the lives of the horses and the work of the scientists.

## MATERIALS

- 4 Pringles cans (empty and clean)
- 4 brown-paper lunch bags
- Plaster of paris (see safety data sheet information in Appendix 4)
- Water
- Mixing bowl
- Spoon or whisk
- Cooking spray
- Permanent marker
- Masking tape
- Rosemary
- Coffee grounds
- Mustard seed
- Elbow macaroni
- Wagon-wheel pasta
- Silver glitter
- Sunflower seeds (in the shell)
- Plastic pony beads (any color)
- Black, brown, and red tempera paint
- Sand
- 5 ft. piece of aluminum or vinyl gutter
- Rice
- Soil (preferably clay)

## SUPPORTING DOCUMENTS

- "Core Sample Instructions" (for teacher use)
- "Events in Earth's History Cards"
- "Reassembled Core Sample Answer Key" (for teacher use)
- "Core Sample Observations" sheet
- "'Understanding Geologic Time' Notes"
- "Geologic Time Explanatory Report Planner"

## SAFETY CONSIDERATIONS

- Review general lab safety protocols.
- Make sure students have appropriate procedures in place for cleanup and disposal of materials from the investigation.
- Remind students that food used in investigations is not to be eaten.
- Tell students to use caution when handling the piece of gutter. To reduce the risk of puncturing skin, place duct tape over the ends of the gutter.
- Have students wash their hands with soap and water after completing the investigation.

## LEARNING-CYCLE INQUIRY

### Engage

In this phase, students will begin to consider the various ways we measure and describe time. They will conclude that larger durations of time are best measured in larger units.

1. Ask students to think-pair-share about the units that we use to measure and talk about time. Record their responses on the board. The list will likely include seconds, minutes, hours, days, weeks, months, and years. Students may also mention decades and centuries. (*Note:* For an explanation of the think-pair-share strategy, see page 346 in Appendix 2.)
2. Conduct a class discussion and ask students to consider which units are best for measuring various lengths of time. In this discussion, lead students to the conclusion that the longer the duration of an event, the larger the unit of time used to describe it. Sample questions include the following:

    - Why do we have so many different units for time?
    - What unit of time would you use to describe a science class?
    - Would you use the same unit of time to describe the amount of time you are at school each day? Why or why not?
    - Why do we typically not use the unit of days to refer to a person's lifetime?
    - How does the length of an event determine what unit of time to use?

3. Share with students that scientists have determined that Earth is approximately 4.6 billion years old. Write this number on the board in word form (4.6 billion) and in standard form (4,600,000,000 years old). To help students appreciate the magnitude of this number, add a few more ages for comparison. You can even make the last age your students' age. Be sure to line up the numbers by the ones digit, as shown in Figure 14.3.

4. Explain that because 4.6 billion is such a large number, scientists break Earth's history into smaller time periods. Ask students to quick write about the following questions:
    - What unit of time might scientists use to describe periods of Earth's history?
    - How might scientists determine how long each period lasts?
5. Collect quick writes and allow students to share their thinking. Introduce the unit question, *How can we use geologic evidence to represent Earth's history?* (*Note:* For an explanation of the quick-write strategy, see page 344 in Appendix 2.)

**FIGURE 14.3. SAMPLE INSTRUCTION BOARD**

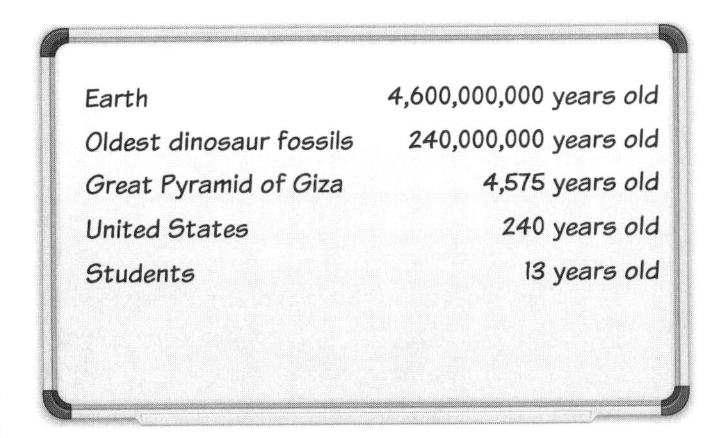


***Assess this phase:*** Only formative assessment is used in this phase, and it comes from two sources: student participation in class discussions and the quick writes. At this point, students should have more questions than correct answers, and they may have no knowledge of geologic time at all. Note the kinds of questions they ask, any relevant information shared in the quick write, and any misconceptions expressed. Subsequent phases offer time to address those misconceptions.

## Explore

In this phase, students observe simulated core samples and create a diagram showing the rock strata found in the samples. They read a list of major geologic and biological events in Earth's past and use evidence from their investigation to assign events to each layer of rock found in the core sample. Finally, students read about the geologic time scale and compare their predictions to the actual time scale.

***Advance preparation:*** First, you need to create the core samples. You will create four model rock cores that can be pieced together to form one continuous rock sample. These rock cores will be created by filling empty Pringles cans with layers of plaster of paris (see recipe). Each can should have five layers of plaster of paris with various ingredients (rosemary, coffee grounds, glitter, etc.) mixed in to represent fossils. These layers will be added in specific order. For detailed instructions, see the supporting document "Core Sample Instructions."

## RECIPE FOR PLASTER OF PARIS

### Ingredients

- Plaster of paris
- Water
- Tempera paint or a dry ingredient to mix into the plaster of paris (use one per batch)

### Notes

- The general ratio of powder to water is 2:1. This ratio may vary depending on the brand of plaster of paris, the water temperature, and the addition of a dry ingredient to the powder.
- Plaster of paris powder is extremely light and fine and, thus, easily dispersed through the air. Avoid getting the powder into your eyes and nose. Wearing a dust mask is highly recommended.

### Preparation

1. Place the plaster of paris in a large mixing bowl and stir to break up any clumps. If you are using a dry ingredient (rosemary, coffee grounds, etc.) stir or whisk it in.
2. Do not add paint to the dry mixture! Add the water to a second container that is large enough to accommodate the amount of paint you plan to mix.
3. Start adding the plaster of paris powder to the water in your mixing container by sprinkling or sifting the powder over the water. Do not add the powder in one spot. Distribute it evenly across the entire surface of the water
4. Tap the sides of the container. Doing so disperses the plaster of paris into the water and removes air bubbles.
5. Continue sprinkling the plaster of paris powder and tapping the sides of the container. Stop tapping when you notice that the powder is almost covering the surface of the water and is not as easily absorbed by the water. Normally, this point would be around the 2:1 ratio mark but may vary slightly.
6. Gently stir the mixture until it reaches a uniform consistency.
7. If using paint, add a few drops in and mix thoroughly.
8. Allow the mixture to stand for one minute before pouring it into the prepared can.

## INSTRUCTIONS FOR CREATING LAYERS

To create the layers in each sample, first label the outside of a clean and empty Pringles can with the appropriate label (see supporting documents). Next, coat the inside of the can with cooking spray. Then, add the first (bottom) layer to the can. The plaster of paris mixture hardens quickly, so prepare the can and have all materials at hand before you begin to mix! Pour the plaster of paris in as soon as it is well mixed. Let each layer dry completely before adding a new layer on top. To save some time, add the bottom layer to each can, let it dry, and repeat this process until all layers have been added. Once the final layer is dry, carefully remove the core from the can. Wrap it in a piece of brown paper (a brown lunch bag that is cut open works well), secure the wrapping with tape, and label the paper with the appropriate core number. Store these carefully until they are needed in class.

Save a small amount of each material that you mixed into the plaster of paris (such as the rosemary, coffee grounds, etc.) so you can set up a key for students in class. Small reusable plastic containers with lids work well for this. Write the name of each material and the type of fossil it represents on index cards to display with the actual materials.

***Advance preparation:*** Before beginning instruction, you must prepare the "Events in Earth's History Cards." Copy and cut apart the cards (see supporting documents), and place the cards for each group in a labeled envelope.

After all preparations have been made, begin the Explore phase. The steps are as follows:

1. Explain to students that to learn about Earth's past, geologists take core samples by drilling down into rock and removing long, tube-shaped samples. Show students a few pictures of rock cores.
2. Explain to students that they will be working in groups to observe one rock core. Distribute one rock core to each group, and instruct students to carefully unwrap the core before beginning to work. Distribute copies of the "Core Sample Observations" handout, and instruct students to sketch the layers (in color) and to record observations in the next column of the handout. Direct students' attention to the samples of each fossil type you have set aside to serve as a key.
3. When groups have finished working, distribute the envelopes containing the "Events in Earth's History Cards." Explain to students that they have been given three major events that happened during the time period represented by their core sample. Students should use evidence from their core sample to identify the layers that correspond to these events. Students should then complete the last two columns on the "Core Sample Observations" handout.
4. Gather the class and explain that their core samples are actually pieces from a larger core. Challenge students to examine the core samples and determine which is the youngest and which is the oldest. Students should be able to easily identify the youngest layer because of the soil layer on top, and they should be able to use matching layers to piece together the entire core.

5. When students have decided, ask the group with the youngest sample to come up and gently place their sample into the section of gutter. The gutter will serve as a holder for the core sample, making it easy for students to refer to it throughout the unit. Designate one end of the gutter as the top. Students should place the soil end (the youngest layer) at the top end of the gutter. Have students place their event cards on the table next to the appropriate layers in their sample.
6. Ask students which groups' sample should be next, and repeat the process until the entire core has been pieced together and all event cards have been added. Take a picture of the finished product so that students can refer to it, as needed, during the rest of the unit.
7. Ask students to identify the oldest part of the core. Beginning with the oldest sample, read the event cards aloud from oldest to youngest. You may also wish to list them on the board, with the oldest event at the bottom and the youngest event at the top.
8. Ask students to do a quick write about the process of observing their core sample, matching events to layers, and piecing the core together. The following questions may be helpful to ask students as they write:
    - What was the hardest part of the investigation? What was the easiest?
    - What did you notice when you tried to match the event cards to the layers?
    - What evidence was most helpful in matching the events to the layers?
    - What did you notice when you put together the entire core sample?
    - Where in the core were the oldest layers? Where were the youngest layers?
    - What did you notice when you read the events from oldest to youngest?
9. Conduct a brief class discussion about students' quick writes. During this discussion, lead students to conclude that fossil evidence was a useful guide for matching events to layers and that, overall, life forms became more complex over time.

QR CODE 3

10. Explain to students that they created a model of the geologic time scale, or a timeline that shows the events in Earth's history. Show students an example of the geologic time scale such as this one from Encyclopedia Britannica Kids (*http://bit.ly/23Zvoj8*; QR code 3). Ask students to think-pair-share about this time scale, using questions such as the following:
    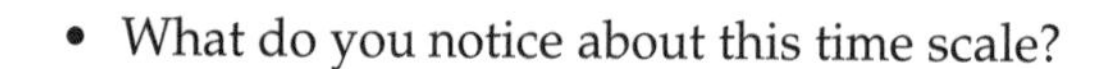
    - What do you notice about this time scale?
    - How does this time scale compare to the one we created in class?
11. Briefly discuss students' reactions and impressions of the time scale.
12. Explain to students that they will be using an interactive module ("Understanding Geologic Time") to learn more about the geologic time scale and how scientists created

it. Students will work individually or in pairs through the module. As they work through the module, students should add notes to the "'Understanding Geologic Time' Notes" graphic organizer. Circulate around the room while students work to answer questions and provide support.

13. Conduct a brief class discussion about the big ideas, student notes, and student questions from the "Understanding Geologic Time" module.

14. Explain to students that they will now view an artist's representation of the geologic time scale. Display the image, "Ages of Rock," found at *http://bit.ly/1O5HrQ9* (QR code 4). Invite students to study the piece, and then complete a quick write on the following questions:

QR CODE 4

   - What is the artist trying to show in this piece?
   - Why did the artist divide the mountain into two halves? What does each half show?
   - What do the people on the top of the mountain represent?
   - How does this piece compare to what you have learned in this unit?

Conclude this phase by asking for volunteers to share their quick writes.

***Assess this phase:*** Formative assessment comes from student participation in the core analysis activity, written observations, and quick writes. Students should be developing an understanding of some of the major events in Earth's history. They should also understand that the fossil record provides evidence of these events and that scientists used the fossil record to create the geologic time scale. It is important to note that "Understanding Geologic Time" introduces students to some concepts, such as radiometric dating and the names of time periods, which are beyond grade-level expectations for middle school students. Anticipate questions and structure discussion so that students grasp the big ideas without trying to memorize details.

## Explain

In this phase, students will write an explanatory piece in which they synthesize evidence from their investigation and reading to explain how rock strata and fossil evidence are used to organize Earth's history into time periods. Review the unit question with students: *How can we use geologic evidence to represent Earth's history?*

1. Inform students that they will write an explanatory report to answer the unit question. Distribute the "Geologic Time Explanatory Report Planner" handout to students. Review the requirements for the explanatory report: introductory statement, details and examples, an illustration, and a conclusion statement. Students should have access to all unit materials as they complete the planner. Review completed

documents to ensure that students understand the requirements before allowing them to write their final copy.

2. Instruct studenst to write their explanatory reports. Students should have access to all unit materials as they complete the report.

*Assess this phase:* The report planner serves as a final formative assessment for this unit. It is essential that students demonstrate an understanding of concepts before writing their actual reports. If students do not show mastery of concepts, return to the Explore phase and re-teach before continuing with the unit.

### Expand

In this phase, students will begin an investigation into the evolution of horses. The activities listed below are identical to the Engage phase in Chapter 15, "The Toes and Teeth of Horses." We strongly recommend teaching these units in sequence by using the Expand phase or the Engage phase to link two learning cycles together.

1. Begin by asking students to share what they know about horses. List the characteristics they mention. Sort the list into physical characteristics and behavioral characteristics.
2. Read excerpts from *Wild Horse Scientists* that will interest your students. Finish with pages 20–25, ending with "They are ancient wanderers who somehow—against all odds—found their way home again."
3. Project the "Horse Evolution" graphic found at *http://bit.ly/1QDjsJS* (QR code 5). Ask students to examine the modern horse and its ancestors. After a few minutes, ask students to describe any differences they notice among the various horses. Record their responses next to the characteristics of horses list compiled earlier. (Most noticeable is the size and number of toes.)

QR CODE 5

4. State that horses have changed over time; they have evolved. Read aloud page 7 of *Evolution: How We and All Living Things Came to Be.*
5. Introduce the question, *In what ways have environmental pressures led to changes in horses over time?*

*Assess this phase:* Assessment in this phase is formative and anecdotal. Monitor student engagement and participation in the class discussion. Pay close attention to their comments and questions, because these may reveal misconceptions about evolution.

## REFERENCES

Dahl, J., S. W. Anderson, and J. C. Libarkin. 2005. Digging into Earth science: Alternative conceptions held by K–12 teachers. *Journal of Science Education* 6 (2): 65–68.

Libarkin, J. C., and S. W. Anderson. 2005. Assessment of learning in entry-level geoscience courses: Results from the Geoscience Concept Inventory. *Journal of Geoscience Education* 53 (4): 394–401.

Libarkin, J. C., S. W. Anderson, J. Dahl, M. Beilfuss, W. Boone, and J. Kurdziel. 2005. Qualitative analysis of college students' ideas about Earth: Interviews and open-ended questionnaires. *Journal of Geoscience Education* 53 (1): 17–26.

National Governors Association Center for Best Practices and Council of Chief State School Officers (NGAC and CCSSO). 2010. *Common core state standards.* Washington, DC: NGAC and CCSSO.

NGSS Lead States. 2013. *Next Generation Science Standards: For states, by states.* Washington, DC: National Academies Press. *www.nextgenscience.org/next-generation-science-standards.*

Rosen, J. *Science*. 2015. Scientists may have found the earliest evidence of life on Earth. October 19. *www.sciencemag.org/news/2015/10/scientists-may-have-found-earliest-evidence-life-earth.*

Name_______________________ Date_______________

# Core Sample Instructions

## CORE SAMPLE A

*Note:* Layers are not drawn to a specified scale. You may vary the thicknesses of the layers within the core sample. The entire sample should be approximately 10 in. long (which is the length of the Pringles can).

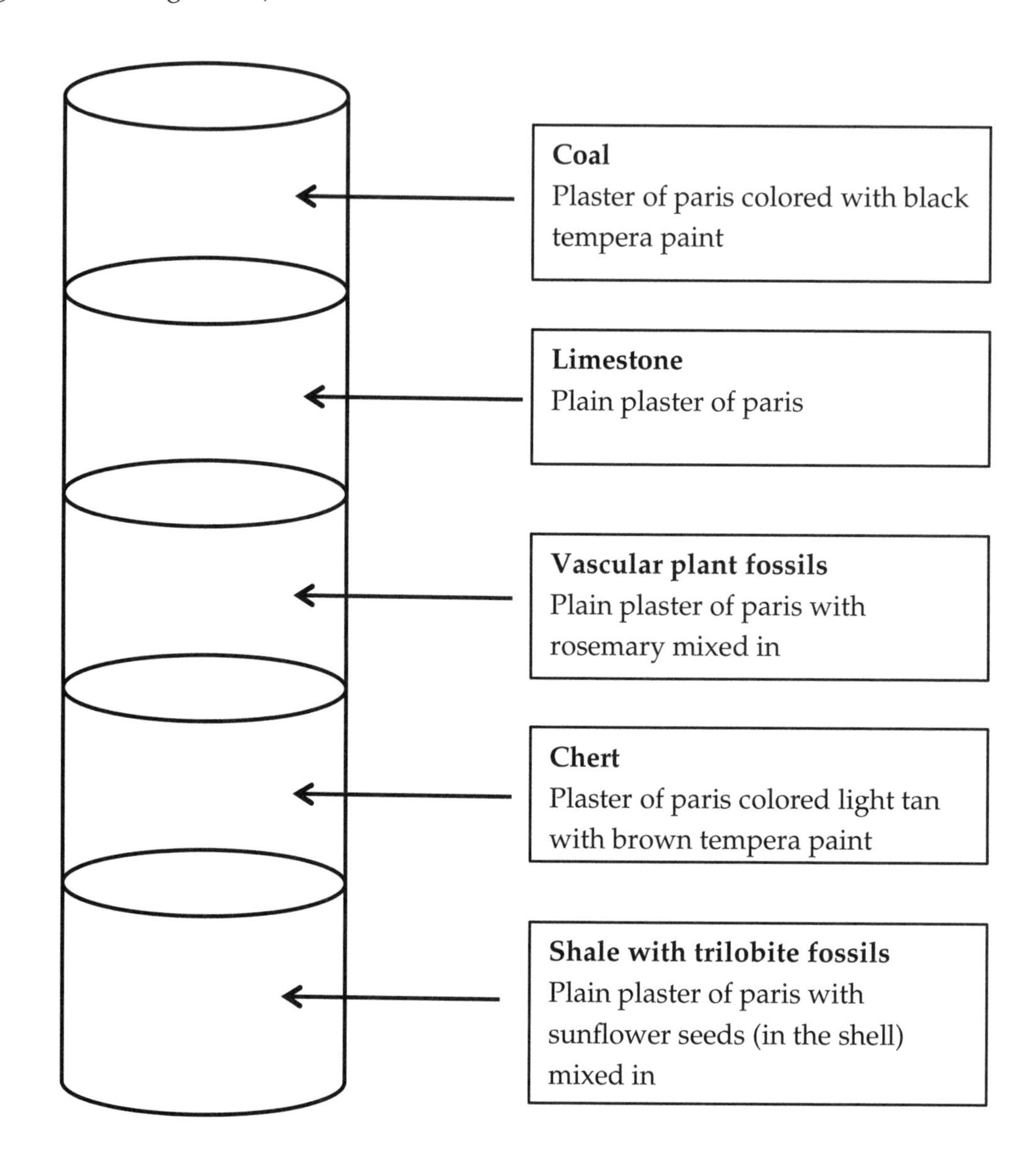

Name______________________________ Date________________

Core Sample Instructions (*continued*)

## CORE SAMPLE B

*Note:* With the exception of the iridium layer, layers are not drawn to a specified scale. Feel free to vary the thicknesses of the layers within the core sample. The entire sample should be approximately 10 in. long (which is the length of the Pringles can).

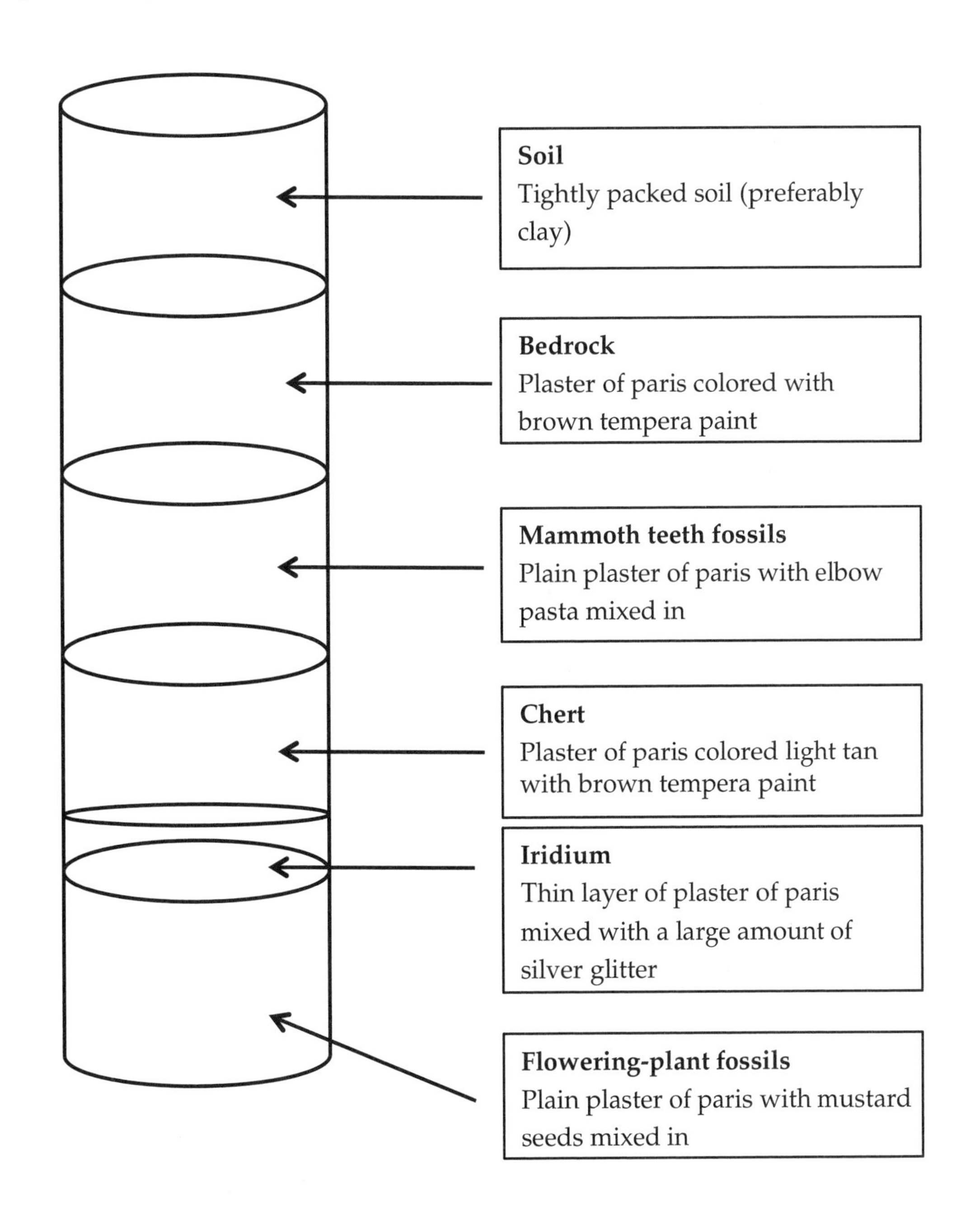

Name______________________________ Date______________

Core Sample Instructions (*continued*)

## CORE SAMPLE C

*Note:* Layers are not drawn to a specified scale. You may vary the thicknesses of the layers within the core sample. The entire sample should be approximately 10 in. long (which is the length of the Pringles can).

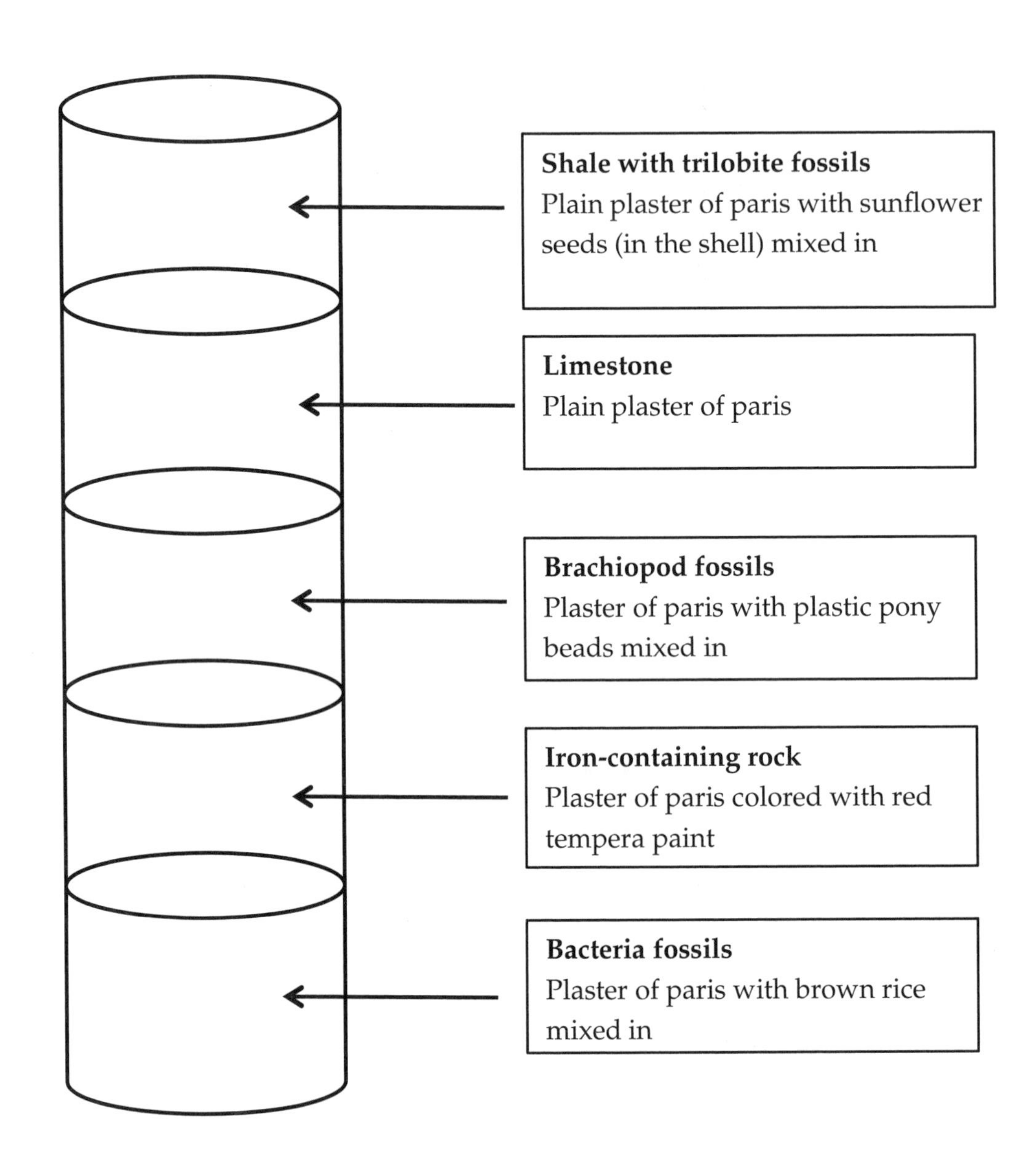

Name_______________________________________ Date_____________________

Core Sample Instructions (*continued*)

## CORE SAMPLE D

*Note:* Layers are not drawn to a specified scale. You may vary the thicknesses of the layers within the core sample. The entire sample should be approximately 10 in. long (which is the length of the Pringles can).

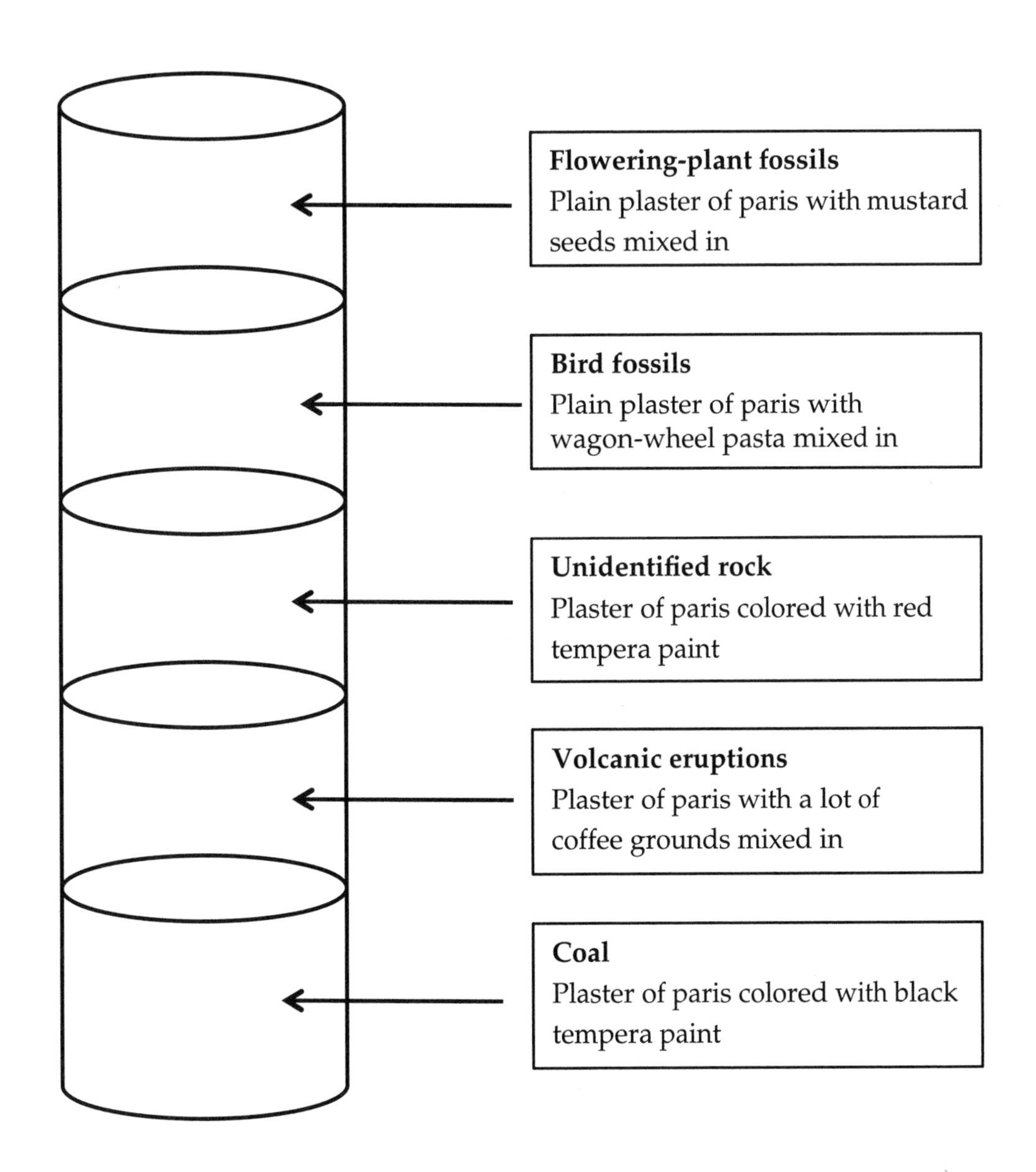

Name______________________ Date______________

# Events in Earth's History Cards

Copy and cut apart these cards, and place each group's cards in a labeled envelope.

| | |
|---|---|
| Core Sample A | The first land plants appeared. |
| Huge plants grew in swamps. The plants' remains would eventually become coal. | Shelled animals called *trilobites* lived in warm ocean waters. |

| | |
|---|---|
| Core Sample B | The first flowering plants appeared. |
| A metorite containing a metal called *iridium* struck Earth. | Wooly mammoths lived in North America and Asia. |

Name______________________________ Date__________________

**Events in Earth's History Cards** (*continued*)

| | |
|---|---|
| Core Sample C | Shelled mollusks called *brachiopods* lived in the ocean. |
| Single-celled bacteria were the only life on Earth. | Shelled animals called *trilobites* lived in warm ocean waters. |

| | |
|---|---|
| Core Sample D | The first flowering plants appeared. |
| Massive volcanic eruptions caused 90% of the world's species to become extinct. | The first birds appeared. Dinosaurs roamed Earth. |

Name________________________ Date____________

# Reassembled Core Sample
# Answer Key

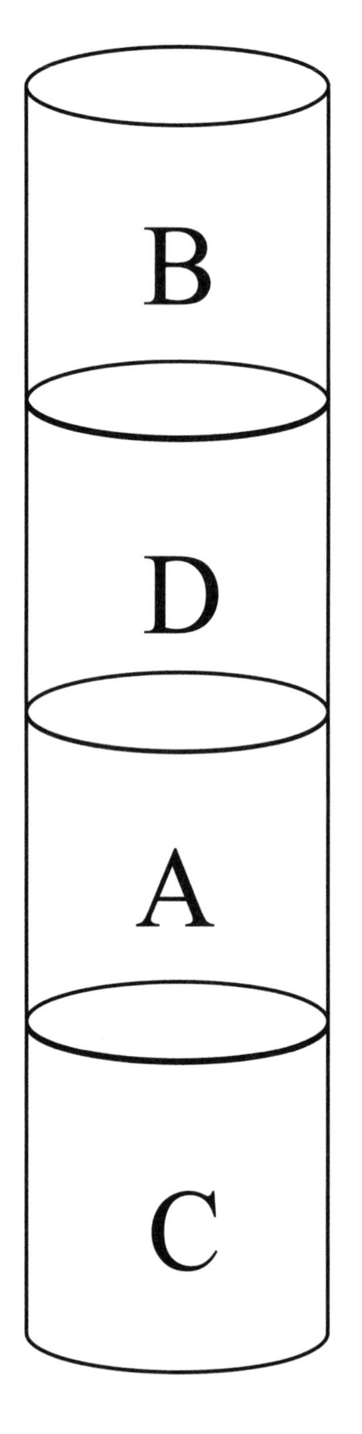


Youngest

Oldest

Name________________________ Date________________

# Core Sample Observations

| Sketch | Observations (Thickness of Layer, Color, Presence of Fossils) | Event in Earth's History | Evidence |
|---|---|---|---|
| | | | |
| | | | |
| | | | |
| | | | |
| | | | |

Name______________________ Date______________

# "Understanding Geologic Time" Notes

As you work through the online module, please add notes for each big idea in the table below. You can also write down any questions you have about what you read.

| Big Idea | Notes | Questions |
|---|---|---|
| 1. Earth has a long, unique history. | | |
| 2. Earth is billions of years old and many things have happened during that time. | | |
| 3. Putting events in order is important. | | |
| 4. We can order events along a timeline. | | |

Name____________________ Date____________

## "Understanding Geologic Time" Notes *(continued)*

| Big Idea | Notes | Questions |
|---|---|---|
| 5. The history of life also has an order. | | |
| 6. Relative time is ordered in rocks. | | |
| 7. Evidence of the events in Earth's history is found within the rocks. | | |
| 8. Radiometric dating gives absolute dates. | | |
| 9. The geologic time scale is an important vertical timeline. | | |

Name_______________________________________ Date_______________________

# Geologic Time Explanatory Report Planner

In the organizer below, plan your report. The report should answer the following question: How can we use geologic evidence to represent Earth's history?

| Introduction | |
|---|---|
| Supporting details and examples | Illustration (describe or quick sketch) |
| Conclusion | |

## Chapter 15

# The Toes and Teeth of Horses

## OVERVIEW

This unit uses nonfiction texts and images to answer the question, *In what ways have environmental pressures led to changes in horses over time?* The investigation focuses on two morphological characteristics, teeth and toes. This unit differs from other units in this book in that the evidence students use to gain knowledge and draw conclusions is entirely secondhand.

This unit assumes that students know and can demonstrate that stratigraphy provides us with evidence of how life on Earth has changed through geologic time. They should also understand that fossils appearing in different strata provide information about what the environment was like during that period and provide evidence of how organisms have changed throughout geologic time. Additionally, students should know that mutations can be beneficial to an organism.

This unit works very well with two other units in this book: "The Genetic Game of Life" (Chapter 7) and "Getting to Know Geologic Time" (Chapter 14). If students have worked through these units, they will be well prepared for "The Toes and Teeth of Horses."

## OBJECTIVES

- Recognize that secondhand evidence can be used to answer scientific questions.
- Describe how organisms change over time in response to environmental pressures.
- Compare anatomical differences between modern and ancestral organisms.
- Communicate science knowledge as narrative nonfiction.

## STANDARDS ALIGNMENT

### *Next Generation Science Standards* (NGSS Lead States 2013)

#### LS4.A: EVIDENCE OF COMMON ANCESTRY AND DIVERSITY

- Anatomical similarities and differences between various organisms living today and between them and organisms in the fossil record, enable the reconstruction of evolutionary history and the inference of lines of evolutionary descent. (MS-LS4-2)

### Common Core State Standards, English Language Arts (NGAC and CCSSO 2010)

GRADES 6–12 LITERACY IN HISTORY/SOCIAL STUDIES, SCIENCE, AND TECHNICAL SUBJECTS

- CCSS.ELA-LITERACY.WHST.6-8.2: Write informative/explanatory texts, including the narration of historical events, scientific procedures/experiments, or technical processes.
- CCSS.ELA-LITERACY.WHST.6-8.9: Draw evidence from informational texts to support analysis, reflection, and research.

## TIME FRAME

- Four to five 45-minute class periods

## SCIENTIFIC BACKGROUND INFORMATION

*Evolution* is often defined as the change in organisms over time. This definition is acceptable, but incomplete. It fails to convey that evolution is a central theory in biology; that evolution has led to the incredible diversity of life on Earth; that all life on Earth shares common ancestors; and that evolution occurs over many, many generations and hundreds of thousands or, perhaps, millions of years. A more complete definition would be that evolution is the central theory in biology that holds that the diversity of life has resulted in changes to common ancestors over a long period of time (National Center for Science Education 2008).

Evidence for evolution comes from multiple sources: the fossil record, homologies (structures derived from a common ancestor), radiometric dating, geographic distribution of plants and animals, and molecular biology. The evidence has been accumulating for more than 150 years and continues to grow. In teaching middle school students about evolution, it is important to use compelling evidence that they can make sense of given their nascent understanding of geologic time, comparative anatomy, and the fossil record. Therefore, in this unit, students explore the evolution of horses by using fossil data representing a relatively short geologic time, examining morphological differences, and studying an organism that they are likely to be familiar with. The scientific background presented here is specifically about the evolution of horses.

*Ungulates*—or hooved, plant-eating mammals—first appear in the fossil record near the end of the Paleocene. Ungulates are divided into two major groups: artiodactyls (e.g., deer, cattle, and sheep) and perissodactyls (e.g., horses, tapirs, and rhinoceroses). Modern-day artiodactyls include more than 100 species; perissodactyls include just 17. Because this unit focuses on horses, we have opted to use the common term *horse* throughout the unit rather than refer to them as perissodactyls.

Artiodactyls and perissodactyls have evolved digitigrade locomotion; that is, they walk on their toes. Artiodactyls, or cloven-hooved mammals, walk on two toes. Perissodactyls walk on one toe (horses) or three toes (rhinos and tapirs). Artiodactyls and perissodactyls' features have evolved to adapt to living in grasslands. Most notably, their teeth adapted in order to better eat grass, and their legs grew longer for faster running. Walking on their toes effectively increases the length of the leg.

Two features of horses that evolved as the environment changed were teeth and toes. In North America at the end of the Paleocene, when the earliest ancestors of horses first appeared in the fossil record, the environment was very warm and very wet. Temperatures are thought to have been about 30°C, with relatively low temperature gradients from pole to pole. Rates of precipitation were very high, and Earth had little ice (University of California Museum of Paleontology 2016). Swampy forests were abundant. By the late Eocene, roughly 20 million years later, the temperature had cooled, Earth had a much greater seasonal range, and the climate was drier. The number of forests had diminished and open prairie–like environments were common. This change in environmental conditions favored organisms with adaptive features that allowed them to survive and reproduce.

One of those features was digitigrade locomotion. At the end of the Paleocene, walking with the pads of four or five toes touching the ground was advantageous. It increased the surface area of the foot that was in contact with the soft, muddy land. The pads believed to have been on the soles of the feet and bottoms of the toes would have increased friction, allowing ancestral horses to walk with greater stability. As the climate became drier and cooler, the land also dried. The length of the "middle" toe of ancestral horses grew longer, and the length of the remaining toes greatly reduced. This adaption essentially lengthened the leg of the ancestral horse, creating a longer stride and greater speed. It also allowed ancestral horses to escape predators in the more open prairie–like environment. Walking with one toe touching the ground allowed ancestral horses to move more quickly in a drier environment with relatively hard-packed land (Levin 1996; Stearn and Carroll 1989).

As mentioned, another feature that evolved as the environment changed was the teeth of horses. Ancestral horses that lived during the late Paleocene and through to the early Eocene ate vegetation such as leaves, fruits, and flowers. As the environment changed, grass emerged as the dominant vegetation. Consequently, the diet of horses changed. On the surface, this diet adjustment may not seem like a big shift because the diets consisted of plant matter in both environments. However, grasses are structurally different from leaves, fruits, and flowers. They contain microscopic silica bodies that grind away at teeth. As the climate continued to change, grasses evolved to contain more silica bodies. Horses that evolved taller teeth were able to eat the high-silica grass and survive (Clarkson 2012).

It is important to remember that changes in the environment did not *cause* the changes in the horses' features. To think otherwise would be inaccurate. The features already existed in the gene pool of the horses, but they were expressed in very low frequency because there was no evolutionary benefit of the feature before the environment changed. Changes in the environment resulted in an increase in the frequency in which the feature was expressed because the feature was an evolutionary benefit.

For more information about stratigraphy and geologic time, see "Getting to Know Geologic Time" (Chapter 14).

## MISCONCEPTIONS

Misconceptions about evolution abound. Many adults and children have misconceptions about evolution, including about how long it takes, whether it has an endpoint, and its causative agents and mechanisms—just to name a few. Listing the breadth and depth of misconceptions about evolution is well beyond the scope of this book. Several misconceptions are listed in Table 15.1. For a more comprehensive list of misconceptions about evolution, see "Misconceptions About Evolution" at *http://bit.ly/1oGay7X* (University of California Museum of Paleontology 2008; QR code 1).

QR CODE 1

### TABLE 15.1. COMMON MISCONCEPTIONS ABOUT EVOLUTION

| Common Misconception | Scientifically Accurate Concept |
|---|---|
| Evolution is about the origin of life on Earth. | Evolution primarily focuses on how life has changed over time. |
| Evolution leads to improvements in organisms. | Evolution may lead to adaptations that enhance survival and reproduction in a particular environment. When the environment changes, however, an adaptation that was at one time beneficial may now be detrimental. Evolution is not a straight path to a "perfect" organism. No such organism exists. An organism may seem perfectly adapted to its current environment, but it may be poorly adapted to survive environmental changes. |
| Organisms can evolve in a single lifespan. | Evolution occurs over many, many generations and, typically, over a long period of time. |

*Source:* University of California Museum of Paleontology 2008.

## MATERIALS

- Various types of paper
- Drawing tools (e.g., markers and colored pencils)

## SUPPORTING DOCUMENTS

- "Horse Toes" activity sheet
- "Change Over Time: Environmental Pressures and the Evolution of Horses" pages
- "Change Over Time: Sequence of Events" pages
- "Horses' Tale Pre-Writing Organizer"
- "Comparing Horses and Whales Through Geologic Time" graphic organizer

## NONFICTION TEXTS

*Evolution: How We and All Living Things Came to Be* by Daniel Loxton (Toronto, ON: Kids Can Press, 2010); Flesch-Kincaid reading level 7.6, published Lexile level 1060L.

Through the use of relatively short passages written in plain language, coupled with colorful illustrations, the author presents evolution in a way that is clear and scientifically accurate. Common questions and misconceptions about evolution are addressed in a thorough and respectful manner.

"Evolution of Whales Animation," by Smithsonian Institution National Museum of Natural History, 2015, *http://s.si.edu/1bfQn3G.* (QR code 2)

QR CODE 2

This animation illustrates how whales evolved from land dwellers to ocean dwellers.

*Evolution Revolution* by Robert Winston (London: DK Publishing, 2009); published Lexile level 1050L.

This text outlines the history of our understanding of biological diversity. The book is divided into four sections: "The Search for Answers," "Darwin and His Theory," "All in the Genes," and "Evolution in Action." Each section provides information in an entertaining and accurate way.

QR CODE 3

"From Feet to Flippers" by The Field Museum, 2016, *http://bit.ly/1XSoPKQ;* Flesch-Kincaid reading level 8.4. (QR code 3)

This slide show provides information about whale evolution, whale ancestors, and the diversity of modern whales.

"From the Horse's Mouth: Teeth Reveal Evolution" by Wynne Parry, Live Science, 2011, *http://bit.ly/1mFkgGd;* Flesch-Kincaid reading level 10.2. (QR code 4)

QR CODE 4

This article describes how a team of paleodentists looked at the wear patterns on fossilized horse teeth to determine why horses became extinct in North America 10,000 years ago.

QR CODE 5

"Horse Evolution" by University of California Museum of Paleontology, 2016, *http://bit.ly/1QDjsJS.* (QR code 5)

This graphic illustrates different species of horses throughout geologic time. The drawings begin with the earliest appearance of an ancestral horse fossil *Hyracotherium* in the Eocene to the most recent species, *Equus.*

QR CODE 6

"Horseevolution.png" from Wikimedia Commons, 2015, *http://bit.ly/21odBn3.* (QR code 6)

This graphic illustrates the evolution of horses. The horses are aligned in stratigraphic columns. Images of the horse's teeth and toes and information on the horse's height accompany each horse image.

QR CODE 7

"On Your Toes" by American Museum of Natural History, 2016, *http://bit.ly/1R6h2qd;* Flesch-Kincaid reading level 8.8. (QR code 7)

This article recounts how horses' hooves developed over time in response to changes in climate.

QR CODE 8

"Talking Teeth" by American Museum of Natural History, 2016, *http://bit.ly/1SQCSlh;* Flesch-Kincaid reading level 5.1. (QR code 8)

This short article discusses the evolutionary advantages of tall teeth in horses.

"Whale Evolution" by American Museum of Natural History, 2016, *http://bit.ly/1QmVCmj;* Flesch-Kincaid reading level 8.6. (QR code 9)

This article provides an overview of whale evolution.

QR CODE 9

*Wild Horse Scientists* by Kay Frydenborg (Boston, MA: Houghton Mifflin Harcourt Books for Young Readers, 2012); Flesch-Kincaid reading level 8.4, published Lexile level 1210L.

This book documents the work two scientists are doing with the wild horses of Assateague Island National Seashore. Beautiful photographs, descriptive text, and the scientists' handwritten notes offer a glimpse into the lives of the horses and the work of the scientists.

## LEARNING-CYCLE INQUIRY

### Engage

This phase introduces students to the concept of evolution. They will begin an investigation into the evolution of horses.

1. Ask students to share what they know about horses. List the characteristics they mention. Sort the list into physical characteristics and behavioral characteristics.
2. Read aloud excerpts from *Wild Horse Scientists* that will interest your students. Finish with pages 20–25, ending with "They are ancient wanderers who somehow—against all odds—found their way home again."
3. Project the "Horse Evolution" graphic (*http://bit.ly/1QDjsJS;* QR code 5). Ask students to examine the modern horse and its ancestors. After a few minutes, ask students to describe any differences they notice among the various horses. Record their responses next to the characteristics of horses list compiled in step 1. (Most noticeable is the size and number of toes.)
4. State that horses have changed over time; they have evolved. Read aloud page 7 of *Evolution: How We and All Living Things Came to Be.*
5. Introduce the question, *In what ways have environmental pressures led to changes in horses over time?*

QR CODE 5

*Assess this phase:* Assessment in this phase is formative and anecdotal. Monitor student engagement and participation in the class discussion. Pay close attention to their comments and questions, because these may reveal misconceptions about evolution.

### Explore

In this phase, students will investigate how the teeth and the toes of horses have changed over time. They will discover that these changes are in response to changing environments.

1. Distribute a copy of the "Horse Toes" handout to each student. Instruct the students to work independently to arrange their cards in the order they think horse toes would have evolved. Each student should explain their reasoning in the space indicated.
2. After students have arranged the images of horse toes and explained their reasoning, instruct them to join their groups and compare their thinking.
3. Ask students to reconsider how they ordered the horse toes, making any changes they think are necessary and explaining their reasoning in the space indicated.
4. When all students have finished their work, ask two to three groups to share their work and their thinking. Allow other students to ask questions and share how their ordering of the horse toes is similar to or different from that of their classmates.

5. Collect students' completed "Horse Toes" sheet for formative assessment. Review them to check students' reasoning about the changes in horse toes.
6. Ask students where the evidence came from that shows that horse toes have changed over time. If you have conducted the "Getting to Know Geologic Time" unit, students will know that the answer is fossils. If not, read page 8 of *Evolution: How We and All Living Things Came to Be.*
7. Follow step 6 with a conversation about the number and position of the toes on *Hyracotherium* and *Equus.*
   a. Project Figure 15.1. Draw attention to the toes on the two horses. Students have already spent time examining the toes, but they may not have thought about how the number and position of the toes would affect a horse's stance and gait. While projecting the image, ask students how they think the number of toes a horse has might affect its ability to move around in different environments. Allow time for some conversation about the question.

## FIGURE 15.1. *HYRACOTHERIUM* AND *EQUUS*

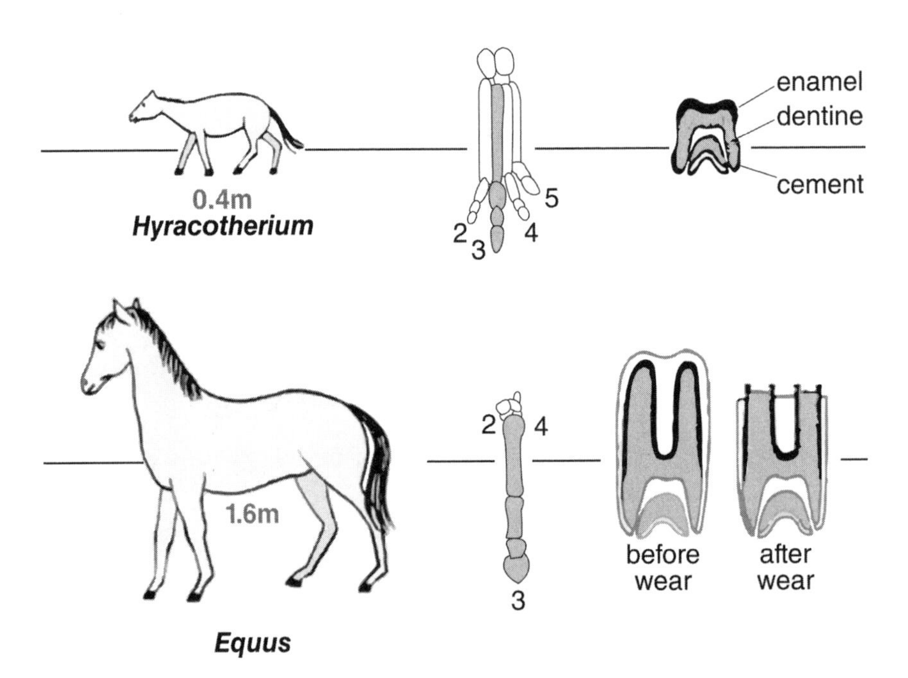


   b. Ask students to stand flat-footed, paying attention to how steady or unsteady they feel. Next, ask students to stand on their tiptoes, paying attention to how steady or unsteady they feel. Ask students to compare how steady or unsteady they felt under the two conditions. Allow a few minutes for discussion.

c. Ask students to think about what it feels like to walk through mud. Tell them to imagine that they are going to walk through mud (but don't have worry about ruining their shoes or getting dirty). Ask them, "Would you be more stable walking through the mud on your tiptoes or on flat feet?" Allow some time for discussion.

d. Look back at the projected image. Ask students which of the horses appears better suited to make their way through swampy, muddy conditions.

8. Project the geologic timeline of horse evolution ("Horseevolution.png"). Note that it is clear from looking at the images that horses changed over time.

9. Ask the class to determine how much time passed between the appearance of *Hyracotherium* and the emergence of *Equus.* This is one of those rare occasions when you are asking for a single correct answer: 50 million years ago. Ask someone to volunteer an answer. Ask the volunteer how he or she determined the answer. Point out (or ask the volunteer to do so) the numbers on the geologic timeline in case some students do not understand how the volunteer arrived at his or answer.

10. Read aloud page 10 of *Evolution: How We and All Living Things Came to Be.* If you have conducted the "Getting to Know Geologic Time" unit, students will be very familiar with the idea of stratigraphy, which will help them connect the two units. If not, this text is a sufficient introduction to stratigraphy for the unit. Ask students to connect this information to the geologic timeline of horse evolution image from step 8. Allow a few minutes for discussion. Establish that 50 million years is a very long time. Close the discussion by noting that long time periods alone are insufficient for evolution to occur.

11. Read aloud page 20 of *Evolution: How We and All Living Things Came to Be.* If you have conducted the "Genetics: The Game of Life" unit, students will be familiar with the concept of mutations. This short reading will help them connect the two units. If not, this short reading will be sufficient for students to identify that genetic changes are a critical part of evolution. If students are completely unfamiliar with the concept of mutations, briefly provide some direct instruction on the topic.

12. Suggest that knowing more about the environment during the past 50 million or so years would be helpful in learning more about the changes in horses over time. Ask students to read the online article "On Your Toes." Students should make notes about the climate and changes in horse toes. Suggest that they organize their notes into a three-column table, as shown in Table 15.2.

### TABLE 15.2. "ON YOUR TOES" NOTES ORGANIZER

| Millions of years ago | Climate and environment | Description of toes |
|---|---|---|
| | | |

13. Provide each student with a copy of the first page of "Change Over Time: Environmental Pressures and the Evolution of Horses." (Do not hand out the Teeth Table on the second page yet.) The opening paragraph describes a sequence of environmental pressures on horses over time. This step is a good opportunity to review with students the textual clues they can use to clarify the sequence of events, specifically signal words such as *before, first, then, soon, after,* and *finally.*
    a. While distributing the copies, tell students not to begin until they are instructed to do so.
    b. After distributing the copies, project the "Change Over Time: Sequence of Events" graphic organizer where all students can see it. You may want to provide copies of the graphic organizers for students. Explain that the graphic organizer is a thinking tool used to keep track of what is happening when you are reading something that has a sequence of events.
    c. Read the paragraph aloud to the class as students follow along. When you come to a signal word (e.g., "horses *first* appeared …") pause, point out the signal word, and record the word where everyone can see it. Continue reading, pointing out and recording signal words as you go.
    d. Before beginning to reread the paragraph, ask students to stop you when you come to a signal word. After students stop you, ask them to summarize the information associated with the signal word. Complete the graphic organizer, sharing your thinking and asking students to share theirs as you go. Leave the completed graphic organizer displayed while students continue to work.
14. Students should first work independently to complete the Toes Table and then rejoin their groups to compare their thinking and come to a consensus. As students are working, circulate around the room, ask students guiding questions, and provide support where needed. Draw the right-hand column where everyone can see it before groups share. Ask two to three groups to share their thinking. As they share, write key points in the appropriate block. Fill in any gaps in the groups' reasoning.
15. Ask students to read one of these online articles: "From the Horse's Mouth: Teeth Reveal Evolution" (above suggested middle school reading level) or "Talking Teeth" (within the suggested middle school reading level). These two articles provide a great opportunity to differentiate instruction and allow for student choice in reading levels. As students read, they should make notes about the climate and changes in horses' teeth and diet. Suggest that they organize their notes into a three-column table as shown in Table 15.2 (p. 319).
16. After students have finished reading their articles, give them a copy of the second page of "Change Over Time: Environmental Pressures and the Evolution of Horses" (the Teeth Table). Repeat parts a–d of step 13; this time, focus on horse teeth.

17. As a class discussion, ask students to use the information they have gathered so far to answer the question, *In what ways have environmental pressures led to changes in horses over time?* Record students' thinking where everyone can see it.
18. Wrap up the Explore phase by asking students to examine pages 82–83 of *Evolution Revolution.* Students should record any new information in each of the tables on the "Change Over Time: Environmental Pressures and the Evolution of Horses" pages. Discuss this new information and respond to questions from the class.

***Assess this phase:*** Assessment in this phase is formative because students are developing their expertise in this phase. Review the "Horse Toes" and "Change Over Time: Environmental Pressures and the Evolution of Horses" pages to check students' reasoning about the changes in horse features over time. During the discussions, listen carefully for misconceptions. By the time students have completed this phase, they should be able to link environmental pressures to evolutionary changes. Pay attention to how students discuss the relationship between the environmental pressures and the changes in horses' features. They may suggest that the changes in the environment *caused* the changes in the horses' features. That conception is inaccurate. Remind students that the features already existed in the gene pool of the horses, but they were expressed in very low frequency because the feature had no evolutionary benefit before the environment changed. Changes in the environment resulted in an increase in the frequency in which the feature was expressed because it now had evolutionary benefits. If students still hang on to the idea of environmental change as a causative agent, reread and discuss page 20 of *Evolution: How We and All Living Things Came to Be.*

## Explain

In this phase, students will write a children's book explaining the 55-million-year history of horses. The book should be in the form of a nonfiction narrative.

1. Tell students that they will be writing a children's book that explains the 55-million-year history of horses. The book must include illustrations that compliment the text. The text must include signal words to help children who read the book better understand the sequence of events. Give each student the "Horses Tale Pre-Writing Organizer." As students are pre-planning, circulate around the room and ask and answer questions. Provide support and guidance where needed. Instruct students to have you check their completed organizer before beginning to write the final product.
2. Offer students a variety of paper and drawing materials. Students may want to access images online, print them, and tape or glue them into their book. We advise against this—not because we are opposed to students using online resources, but because science is expressed in many different ways. The ability to communicate science in a variety of ways is an important part of sharing scientific knowledge. Assure students

that they are not being assessed on their ability to draw horses or landscapes. Explain that they are being assessed on their ability to communicate what they know about the evolution of horses. As in step 1, circulate around the room to ask and answer questions and provide support and guidance where needed.

3. Gather students' finished books and evaluate them. Then, arrange for students to share their books with elementary students in the district, if possible.

***Assess this phase:*** Assessment in this phase is summative. Students should be able to communicate their expertise on horse evolution through the children's book they write.

## Expand

In this phase, students will compare the evolution of whales to the evolution of horses. The activities in this phase focus on reasoning as opposed to arriving at a correct answer.

***Advance preparation:*** Because students will be accessing online resources independently or in small groups, bookmark and organize the websites in advance. Numerous online tools can be used for this task. Provide additional print and online materials about modern-day horses and whales.

After preparations have been made, begin the Expand phase. The steps are as follows:

QR CODE 2

1. Begin by sharing that whales evolved in the same geologic time frame as horses. They experienced the same environmental pressures in terms of climate change, yet they evolved from land dwellers to ocean dwellers. Wonder aloud, "How could it be that two vertebrates—both mammals living on the land—could take such dissimilar evolutionary paths?"
2. As a class, brainstorm possible factors that may have played a role in the evolutionary paths of whales and horses. Record students' ideas where they can be saved for the duration of the activities.

QR CODE 3

3. Distribute copies of "Comparing Horses and Whales Through Geologic Time" to each student. Students can work independently, in pairs, or in small groups to complete the graphic organizer. Direct students to use the electronic and print resources available to them to gather the information they need to answer the question, "How could it be that two vertebrates—both mammals living on the land—could take such dissimilar evolutionary paths?"
   - "Evolution of Whales Animation" (*http://s.si.edu/1bfQn3G*; QR code 2)
   - "From Feet to Flippers" (*http://bit.ly/1XSoPKQ*; QR code 3)
   - "Whale Evolution" (*http://bit.ly/1QmVCmj*; QR code 9)

QR CODE 9

4. When students have finished gathering data, have them share their findings with the class. Then, ask them to share what they think may have contributed to horses and whales taking such dissimilar evolutionary paths.

*Assess this phase:* Assessment of student work in this phase is formative. As students contemplate how the evolutionary lines of horses and whales could have taken such different paths, pay close attention to their reasoning. Students should recognize that even though the environmental pressures were the same with regard to climate, other factors affected the evolutionary paths of horses and whales.

## REFERENCES

Clarkson, N. 2012. Why did horses die out in North America? Horsetalk. *www.horsetalk.co.nz/2012/11/29/why-did-horses-die-out-in-north-america/#axzz41OuWc8cN.*

Levin, H. L. 1996. *The Earth through time.* Orlando, FL: Saunders College Publishing.

National Center for Science Education. 2008. Defining evolution. *http://ncse.com/evolution/education/defining-evolution.*

National Governors Association Center for Best Practices and Council of Chief State School Officers (NGAC and CCSSO). 2010. *Common core state standards.* Washington, DC: NGAC and CCSSO.

NGSS Lead States. 2013. *Next Generation Science Standards: For states, by states.* Washington, DC: National Academies Press. *www.nextgenscience.org/next-generation-science-standards.*

Stearn, C. W., and R. L. Carroll. 1989. *Paleontology: The record of life.* New York: John Wiley & Sons.

University of California Museum of Paleontology. 2008. Misconceptions about evolution. *http://evolution.berkeley.edu/evolibrary/misconceptions_faq.php#a1.*

University of California Museum of Paleontology. 2016. Understanding evolution. *http://evolution.berkeley.edu/evolibrary/home.php.*

Name________________________________ Date________________

# Horse Toes

The toes of horses changed over time as horses adapted to changes in their environments. The images at the bottom illustrate how the bones in the feet of horses changed over time.

## PROCEDURE

1. Cut the images of horse toes at the bottom of the sheet apart.
2. Work by yourself to arrange the images in the order you think the changes may have occurred.
3. Why did you arrange the images this way? List your reasons.

| I arranged the horse toes as I did because |
|---|

4. Join your group and compare how you and your group mates arranged the horse toes.
5. Each person in the group should explain how he or she arranged the horse toes.
6. Are you satisfied with your organization of the horse toes after listening to others in our group? If not, reorganize the horse toes until you are satisfied.
7. In the space below, explain why you rearranged the horse toes as you did.

| I rearranged the horse toes as I did because |
|---|

8. Tape your final arrangement of horse toes to the bottom or back of this sheet.

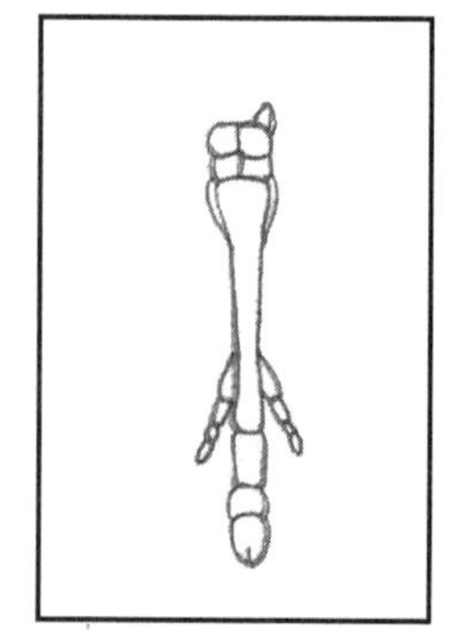
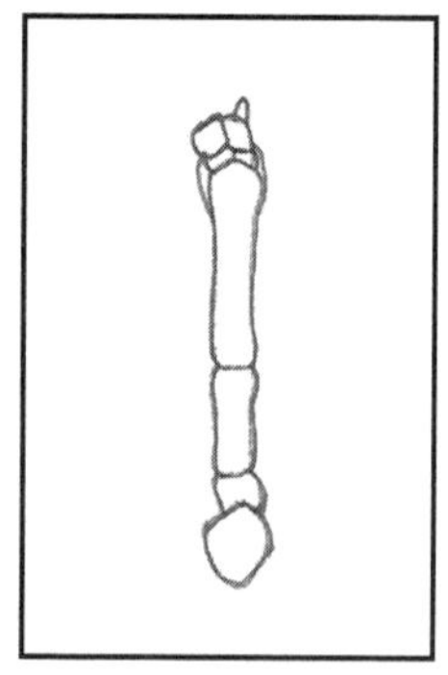
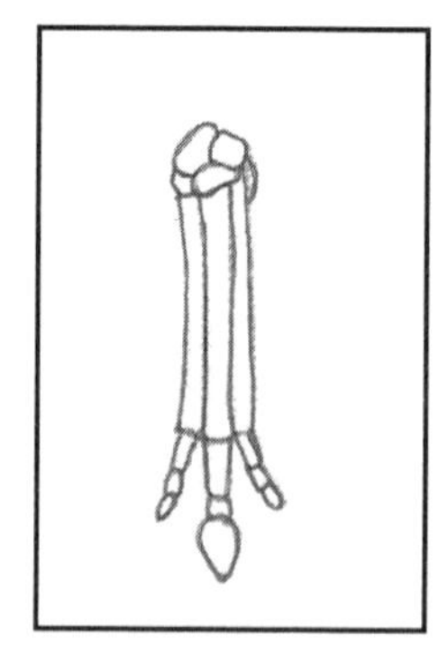
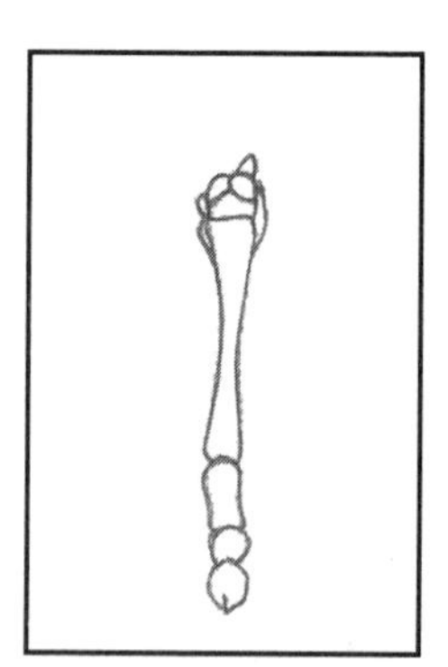
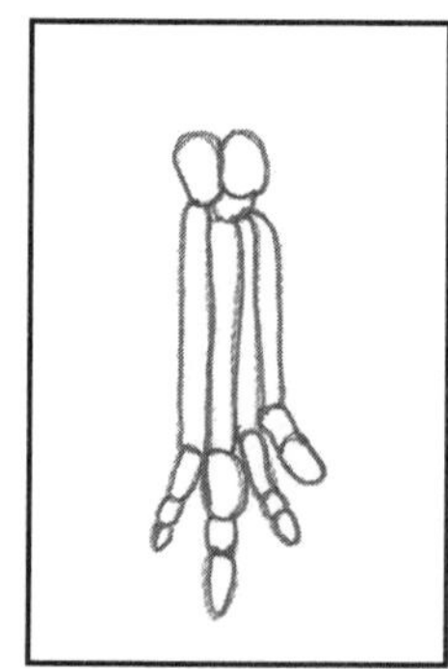

Name________________________________________ Date______________________

# Change Over Time: Environmental Pressures and the Evolution of Horses

It is clear that horses changed over time. But why? What environmental pressures did they face? Early forms of horses first appeared in the fossil record about 55 million years ago in North America. At that time, North America was very warm and moist. The environment was a tropical rainforest with many trees and small plants. The ground was soft and squishy; it was swampy, too. Then, the climate began to cool and, by 35 million years ago, North America was mostly prairie-like grasslands. The grasses were much tougher to eat than the leaves and fruits of the trees and small plants that covered the land millions of years earlier. About 10 million years ago (25 million years after the climate began to change), the climate was cold and harsh. Grasslands covered North America. There were also trees and other plants, but grass was the major food source for herbivores. The grasses contained particles of silica that slowly grind down the teeth. By this time, the ground was hard and easy to run on. The environmental pressures that horses faced over a 45-million-year time span played a very important role in their evolution. Most noticeable are the changes in horses' teeth and toes.

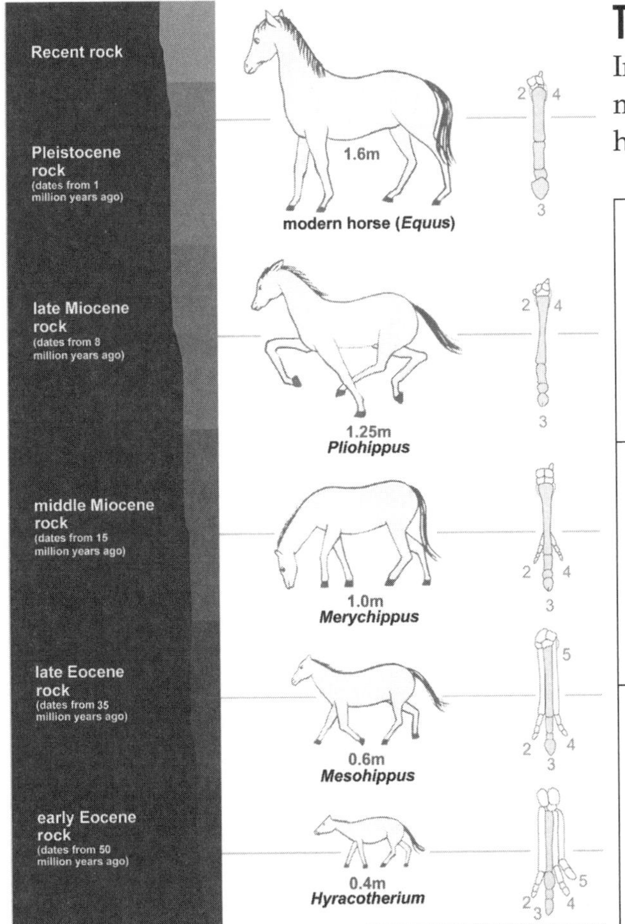


## TOES TABLE

In the column below, write down the environmental conditions, the changes in the **toes** of horses, and the benefits of those changes.

| 10 million years ago |
|---|
| 35 million years ago |
| 50–55 million years ago |

Name________________________________________ Date______________________

## Change Over Time: Environmental Pressures and the Evolution of Horses (*continued*)

### TEETH TABLE

In the column on the right, write down the environmental conditions and available food, the changes in the **teeth** of horses, and the benefits of those changes.

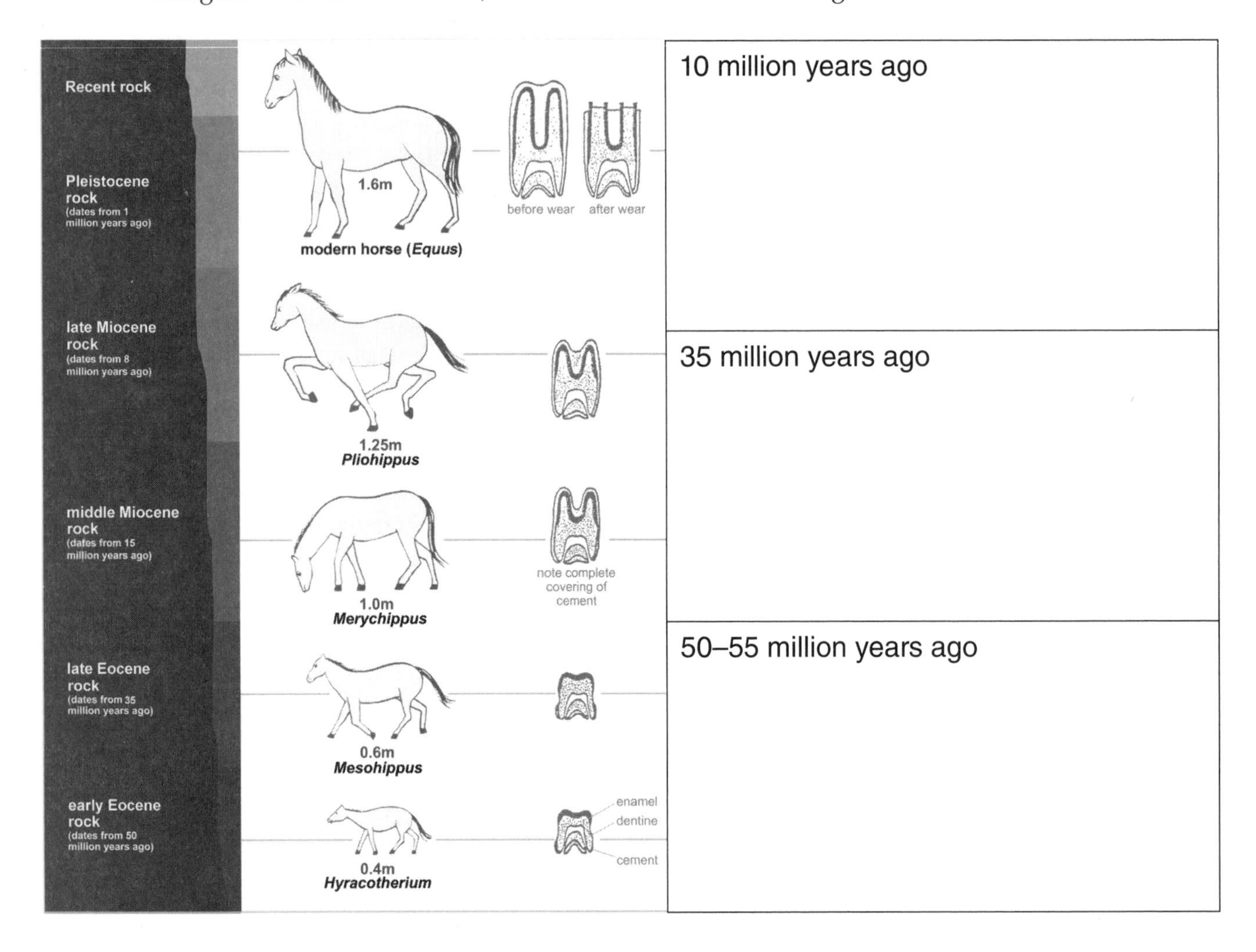

Name________________________ Date________________

# Change Over Time: Sequence of Events

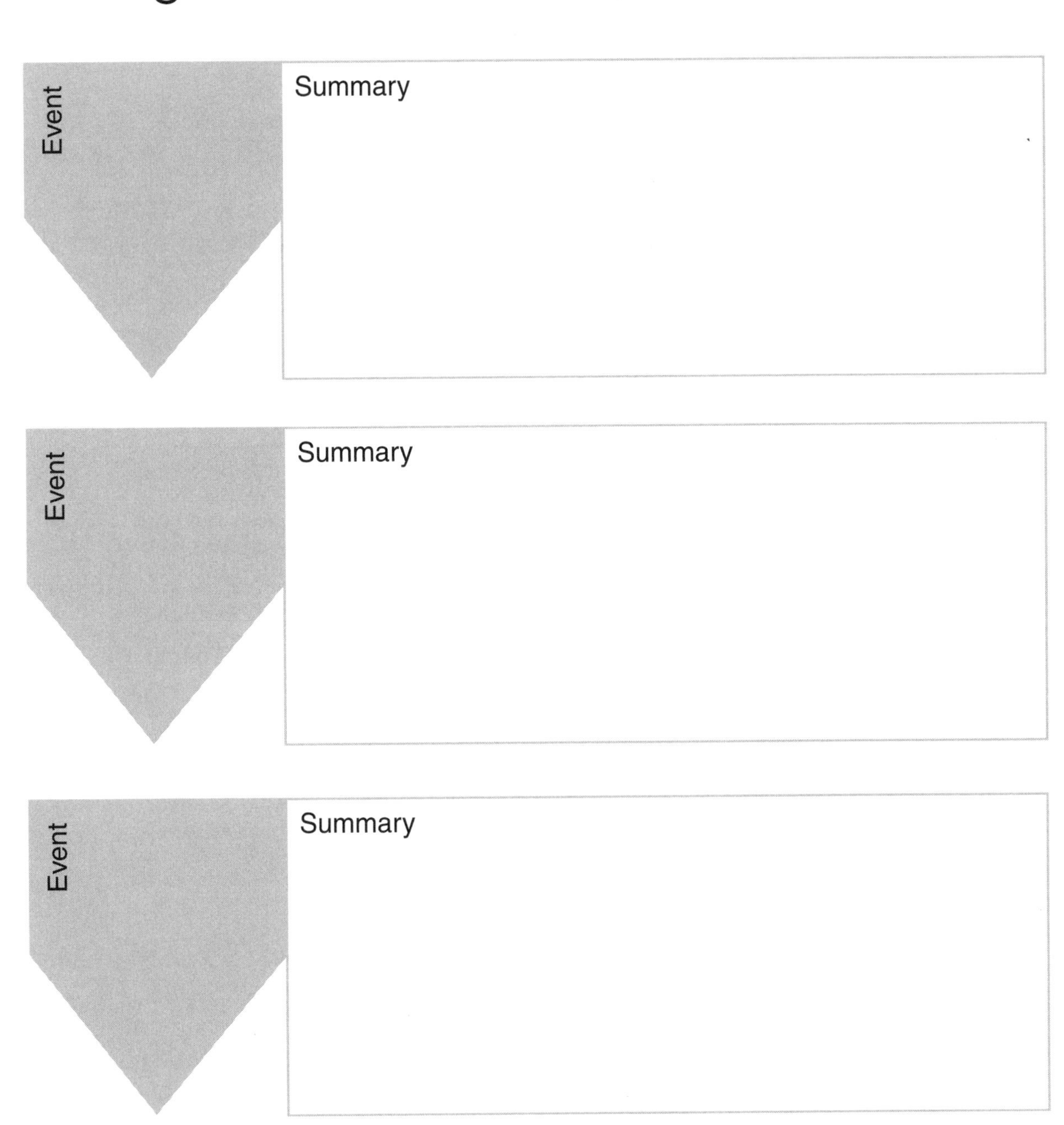

Name____________________ Date______________

# Horses Tale Pre-Writing Organizer

Use this pre-writing organizer to record and organize the information for your children's book. Add more rows as needed.

| Event | Signal word | Text | Image |
| --- | --- | --- | --- |
| | | | |

| Event | Signal word | Text | Image |
| --- | --- | --- | --- |
| | | | |

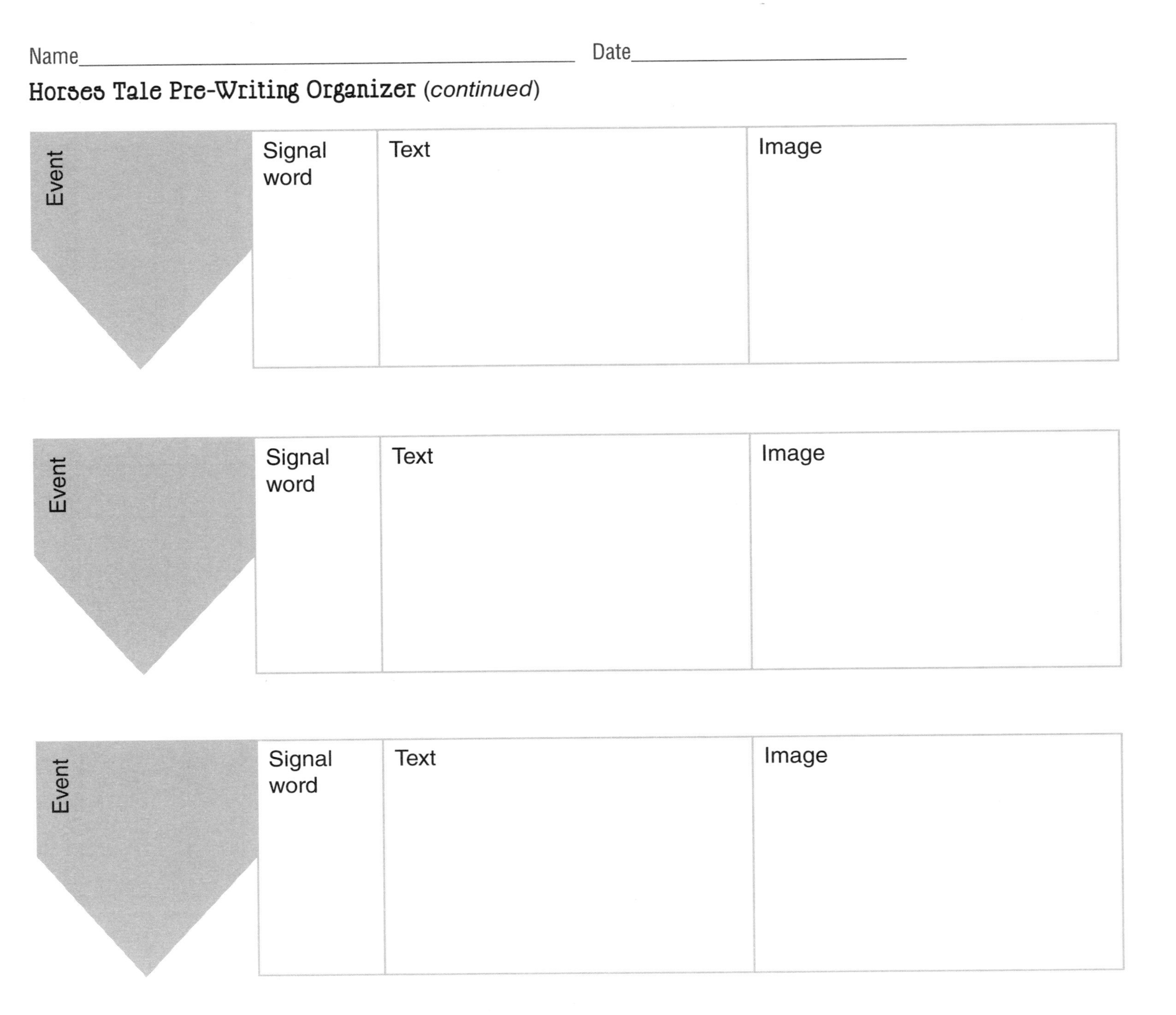
Name
Date
Horses Tale Pre-Writing Organizer (continued)
Event
Signal word
Text
Image
Event
Signal word
Text
Image
Event
Signal word
Text
Image

Name______________________________ Date________________

# Comparing Horses and Whales Through Geologic Time

Think about what horses and whales ate, where they lived on the land, how they moved about, and other characteristics. Then, complete the organizer below.

How were the ancestors of modern horses and whales similar?

**Horses and Whales**

How were the ancestors of modern horses and whales different?

| Ancestral Horses | Ancestral Whales |
| --- | --- |
| | |
| | |
| | |
| | |
| | |
| | |
| | |

Name______________________ Date______________

## Comparing Horses and Whales Through Geologic Time *(continued)*

Think about what horses and whales ate, where they live, how they moved about, and other characteristics. Then, complete the organizer below.

How are modern horses and whales similar?

**Horses and Whales**

How are modern horses and whales different?

| Modern Horses | Modern Whales |
| --- | --- |
| | |
| | |
| | |
| | |
| | |
| | |
| | |

## Appendix 1

# Rubrics and Achievement Grading Standards

The following rubrics can be used to assess the work products generated in the Explain and, occasionally, Expand phases of the inquiry units. You can then compare the completed rubrics with the Achievement Grading Standards (Table A1.1) to determine the percentage to score an assignment. Notice that the lowest score on the Achievement Grading Standards chart is 75%. If a student falls below that level, return to the instructional activities in the unit and re-teach until the student can meet at least some of the criteria on the Science and Literacy rubric (Table A1.2, p. 334).

Following the Achievement Grading Standards chart are two rubrics that can be used to assess students' science process skills (Tables A1.3 and A1.4, pp. 335–339). As discussed in Chapter 2, we believe that process skills, such as observing, inferring, and defining operationally, still have important roles to play in science classrooms. Those skills support the science and engineering practices described in current standards (Table A1.5, p. 340). Although language has changed, the behaviors and habits of mind necessary for science have not.

The first rubric is designed for teacher use; the second rubric is for students to use to self-assess. You can use the rubrics periodically throughout the year to monitor the development of essential skills. Portions of the rubric could also be used before and after direct instruction in a particular skill. For teachers wishing to assess the science and engineering practices directly, one source of teacher-created rubrics is available through the Literacy Design Collaborative. That group also offers a rubric for assessing the practices (*http://bit.ly/1sY1Xjx;* QR code 1).

QR CODE 1

### TABLE A1.1. ACHIEVEMENT GRADING STANDARDS

| Percent (Grade) | Standard |
|---|---|
| 100% | Based on the rubric, the student has exceeded all of the criteria. |
| 95% | Based on the rubric, the student has met all and exceeded many of the criteria. |
| 90% | Based on the rubric, the student has met all and exceeded some of the criteria. |
| 85% | Based on the rubric, the student has met all of the criteria. |
| 80% | Based on the rubric, the student has met most of the criteria. |
| 75% | Based on the rubric, the student has met some of the criteria. |

# Appendix 1

TABLE A1.2. SCIENCE AND LITERACY RUBRIC

| Area | Criteria Exceeded | Criteria Met | Criteria Not Sufficiently Met |
|---|---|---|---|
| Science Content | Content illustrates an accurate and thorough understanding of scientific concepts as defined by the objectives and is able to tie concepts together in a coherent way. | Content illustrates an accurate and thorough understanding of scientific concepts as defined by the objectives. | Content illustrates a limited or inaccurate understanding of scientific concepts as defined by the objectives. |
| Scientific Terminology (vocabulary) | Work products demonstrate a correct, developmentally appropriate understanding of all terminology through writing or speaking and ability to draw relationships between the terms. | Work products demonstrate a correct, developmentally appropriate understanding of all terminology through writing or speaking. | Work products demonstrate a limited or inaccurate understanding of scientific terms. |
| Use of Evidence (from texts and investigations) | Work product explicitly refers to evidence. | Work product consistently alludes to evidence. | Work product is not supported by evidence. |
| Clarity of Message | Writing, visual image use, or speaking is coherent, complete, and conveys sufficient information to elaborate on the points raised. | Writing, visual image use, or speaking is coherent, complete, and conveys sufficient information to make the point. | Writing, visual image use, or speaking lacks coherence, is incomplete, or includes insufficient information. |
| Organization | Organization of the work product enhances the meaning conveyed. | Components of the work product are presented in a logical sequence and arrangement. | Components of the work product are not presented in a logical sequence and arrangement. |
| Format | Work product adheres to the assigned format and includes additional features that enhance the product. | Work product adheres to the assigned format. | Work product does not adhere to the assigned format. |
| Quality of Work | Work product exceeds the student's normal standard of work. | Work product meets the student's normal standard of work. | Work product falls below the student's normal standard of work. |

# Appendix 1

## TABLE A1.3. SCIENCE PROCESS SKILLS RUBRIC

| Skills | Independent | Developing | Beginning |
|---|---|---|---|
| *Basic Science Process Skills* | | | |
| Observing | The student independently makes and records detailed observations using his or her senses and appropriate measuring tools. Recorded observations are clear and free of ambiguous language and illustrations. | With teacher or peer assistance, the student makes and records detailed observations using his or her senses and appropriate measuring tools. Recorded observations are described in clear language and illustrations. | With teacher assistance, the student makes and records simple observations using his or her senses and appropriate measuring tools. Recorded observations may include ambiguous language and illustrations. |
| Measuring | Working independently with appropriate tools, the student makes and records detailed measurements. | Using appropriate tools with teacher or peer assistance, the student makes and records detailed measurements. | Using appropriate tools with teacher assistance, the student makes and records simple measurements. |
| Classifying | Working independently, the student sorts and organizes objects by group or order and documents the process with words and with illustrations, diagrams, or flow charts. | With teacher or peer assistance, the student sorts and organizes objects by group or order and describes the process with words or with illustrations, diagrams, or flow charts. | With teacher assistance, the student sorts and organizes objects by group or order and describes the process with words. |
| Inferring | Working independently, the student clearly explains a possible relationship between previous knowledge and new observations. | With teacher or peer assistance, the student explains a possible relationship between previous knowledge and new observations. | Only with teacher assistance, the student explains a possible relationship between previous knowledge and new observations. |
| Predicting | Working independently, the student makes well-reasoned predictions that are based on previous observations. | With teacher or peer assistance, the student makes thoughtful predictions that are based on previous observations. | Only with teacher assistance, the student makes predictions that are based on previous observations. |
| Communicating | Working independently, the student communicates detailed information with words or diagrams in a format of his or her choosing. | With teacher or peer assistance, the student communicates basic information with words or simple diagrams in a format of his or her choosing. | With teacher assistance, the student communicates basic information with words or simple diagrams in a prescribed format. |

Table A1.3 (*continued*)

| Skills | Independent | Developing | Beginning |
|---|---|---|---|
| *Integrated Science Process Skills* | | | |
| Controlling Variables | Working independently, the student identifies dependent and independent variables. | With teacher or peer assistance, the student identifies dependent and independent variables. | Only with teacher assistance, the student identifies dependent and independent variables. |
| Defining Operationally | Working independently, the student determines how the dependent variable will be measured. | With teacher or peer assistance, the student determines how the dependent variable will be measured. | Only with teacher assistance, the student determines how the dependent variable will be measured. |
| Interpreting Data | Working independently, the student makes evidence-based claims. | With teacher or peer assistance, the student makes evidence-based claims. | Only with teacher assistance, the student makes evidence-based claims. |
| Formulating Hypotheses | Working independently, the student makes a cause-and-effect statement about the expected outcome of the experiment. | With teacher or peer assistance, the student makes a cause-and-effect statement about the expected outcome of the experiment. | Only with teacher assistance, the student makes a cause-and-effect statement about the expected outcome of the experiment. |
| Experimenting | Working independently, the student designs and conducts an experiment. | With teacher or peer assistance, the student designs and conducts an experiment. | With teacher assistance, the student conducts a teacher-led experiment. |
| Formulating Models | Working independently, the student creates an evidence-based representation of a process or concept. | With teacher or peer assistance, the student creates an evidence-based representation of a process or concept. | Only with teacher assistance, the student creates an evidence-based representation of a process or concept. |

# Appendix 1

## TABLE A1.4. MY SCIENTIFIC PROCESSES RUBRIC

Place a check mark beside the statement that tells how you feel about your ability to do each of the following things. Then, for each skill, complete the sentence about how you can strengthen that particular skill.

| | | |
|---|---|---|
| *Observations:* I can make and record detailed observations using my senses and appropriate measuring tools. My observations are clearly described with words and with illustrations or diagrams. | | I need my teacher's help. |
| | | I need my classmate's help. |
| | | I can do this by myself. |
| I can strengthen my skills by | | |

| | | |
|---|---|---|
| *Measuring:* Using appropriate tools, I can make and record detailed measurements. | | I need my teacher's help. |
| | | I need my classmate's help. |
| | | I can do this by myself. |
| I can strengthen my skills by | | |

| | | |
|---|---|---|
| *Classifying:* I can sort and organize objects by group or order, and document the process with words and with illustrations, diagrams, or flow charts. | | I need my teacher's help. |
| | | I need my classmate's help. |
| | | I can do this by myself. |
| I can strengthen my skills by | | |

| | | |
|---|---|---|
| *Inferring:* I can clearly explain a possible relationship between what I already know and new observations. | | I need my teacher's help. |
| | | I need my classmate's help. |
| | | I can do this by myself. |
| I can strengthen my skills by | | |

**Table A1.4** (*continued*)

| *Predicting:* I can make reasonable predictions that are based on previous observations. | | I need my teacher's help. |
|---|---|---|
| | | I need my classmate's help. |
| | | I can do this by myself. |
| I can strengthen my skills by | | |

| *Communicating:* I can share detailed information with words and/or illustrations or diagrams in a format that I pick. | | I need my teacher's help. |
|---|---|---|
| | | I need my classmate's help. |
| | | I can do this by myself. |
| I can strengthen my skills by | | |

| *Controlling variables:* I can identify dependent and independent variables. | | I need my teacher's help. |
|---|---|---|
| | | I need my classmate's help. |
| | | I can do this by myself. |
| I can strengthen my skills by | | |

| *Defining operationally:* I can describe how the dependent variable will be measured. | | I need my teacher's help. |
|---|---|---|
| | | I need my classmate's help. |
| | | I can do this by myself. |
| I can strengthen my skills by | | |

| *Interpreting data:* I can make evidence-based claims. | | I need my teacher's help. |
|---|---|---|
| | | I need my classmate's help. |
| | | I can do this by myself. |
| I can strengthen my skills by | | |

# Appendix 1

| | | |
|---|---|---|
| *Making hypotheses:* I can make a cause-and-effect statement about the expected outcome of the experiment. | | I need my teacher's help. |
| | | I need my classmate's help. |
| | | I can do this by myself. |
| I can strengthen my skills by | | |

| | | |
|---|---|---|
| *Experimenting:* I can design and conduct an experiment. | | I need my teacher's help. |
| | | I need my classmate's help. |
| | | I can do this by myself. |
| I can strengthen my skills by | | |

| | | |
|---|---|---|
| *Making models:* I can make an evidence-based model of a process or concept. | | I need my teacher's help. |
| | | I need my classmate's help. |
| | | I can do this by myself. |
| I can strengthen my skills by | | |

# Appendix 1

**TABLE A1.5. CORRELATION OF SCIENCE AND ENGINEERING PRACTICES AND PROCESS SKILLS**

| Science and Engineering Practices | Supporting Process Skills |
|---|---|
| Asking questions (for science) and defining problems (for engineering) | • Communicating<br>• Observing |
| Developing and using models | • Communicating<br>• Formulating models<br>• Observing<br>• Predicting |
| Planning and carrying out investigations | • Controlling variables<br>• Defining operationally<br>• Experimenting<br>• Formulating hypotheses<br>• Inferring<br>• Measuring<br>• Observing<br>• Predicting |
| Analyzing and interpreting data | • Controlling variables<br>• Defining operationally<br>• Experimenting<br>• Formulating hypotheses<br>• Inferring<br>• Measuring<br>• Observing<br>• Predicting |
| Using mathematics and computational thinking | • Formulating hypotheses<br>• Formulating models<br>• Interpreting data<br>• Predicting |
| Constructing explanations (for science) and designing solutions (for engineering) | • Classifying<br>• Communicating<br>• Formulating models<br>• Inferring<br>• Interpreting data<br>• Observing<br>• Predicting |
| Engaging in argument from evidence | • Communicating<br>• Interpreting data<br>• Inferring<br>• Predicting |
| Obtaining, evaluating, and communicating information | • Communicating<br>• Observing |

**Appendix 2**

# Background Information: Literacy Strategies and Techniques

hroughout our units, we have referenced and included a variety of strategies and techniques intended to support students' literacy development. In this appendix, we provide background information on each of these strategies.

## ANCHOR CHARTS

Anchor charts are typically large sheets of paper used to record content, processes, strategies, guidelines, or cues during instruction. Posting anchor charts in the classroom reminds students of prior learning and allows them to make connections between prior knowledge and new information. Students can also use these charts as tools during independent work. Wendy Seger (as cited in Reinhartz 2015) identifies five essential features of an anchor chart: (1) has a single focus, (2) is co-constructed with the students, (3) has an organized appearance, (4) matches the students' developmental level, and (5) supports ongoing learning. Anchor charts are an effective scaffold for students in all content areas, including reading (Vlach and Burcie 2010), mathematics (Di Teodoro et al. 2011), and science (Reinhartz 2015).

## ARGUMENTATION

Argumentation occurs across academic disciplines, but for the purposes of this book, we focus on scientific argumentation. During an inquiry activity, students should make claims on the basis of observable evidence. They should also be able to justify why the evidence supports the claim, often using accepted terminology and concepts to do so. In some cases, students may also construct a rebuttal in which they present alternative claims and use evidence to disprove those claims. One popular approach to teaching argumentation is the Claim, Evidence, and Reasoning Framework (McNeil and Krajcik 2011).

## CLOSE READING

Close reading is a strategy that promotes a deep understanding of a text. Over the course of multiple readings, students attend to key ideas and supporting details, reflect on the meaning of words and sentences, and focus on how the author develops ideas. Although close reading does not have a set instructional sequence, it is a multi-step process. Students begin by reading the text individually, focusing on the main ideas and details at first. A brief class discussion or think-pair-share (see p. 346) allows the teacher to assess student understanding. A second reading focuses on what are known as the *craft* and *structure* of the text—that is, the author's choice of vocabulary and text structures. Another discussion follows to check for understanding. A third reading asks students to synthesize information and ideas. Following a third reading, students write a journal response to a text-dependent question—one that requires them to refer to the text for evidence.

## FISHBOWL

The fishbowl technique is a strategy that helps students practice being both contributors and listeners during a discussion. In this strategy, students are divided into two groups. One group of students sits in the "fishbowl" circle, where they participate in a discussion, asking questions and sharing information and opinions. The rest of the students are on the outside of the circle, and they listen carefully to the ideas, pay attention to the process, and sometimes take notes. Then, the two groups switch roles. This technique helps to ensure active engagement and participation for all students. It is also helpful to use this strategy when you want to model and practice discussion techniques.

## FRAYER MODEL

The Frayer model (Frayer, Frederick, and Klausmeier 1969) is a vocabulary strategy that asks students to develop their understanding of the concept rather than simply memorize a definition. The model itself is a diagram with the vocabulary term in the center. In boxes around the term, students provide a definition, list characteristics, and provide examples and nonexamples of the concept. When used purposefully and selectively, the Frayer model can provide an opportunity to clarify concepts and deepen understanding of terms and concepts. It should be reserved for rich, complex vocabulary because it consumes more time than other approaches (Greenwood 2004).

## GALLERY WALK

In a gallery walk, students move from station to station to view each other's work, pose questions, and give feedback (Keely 2014). The teacher also circulates the room during this time to view student work and listen in as students discuss the work at hand. Gallery walks capitalize on the benefits of movement during a lesson, foster engagement, and provide a powerful tool for formative assessment.

## GRAPHIC ORGANIZERS

Graphic organizers take many forms. Commonly used graphic organizers include webs, maps, tables, and diagrams. They provide a visual representation of concepts and illustrate how concepts are linked (Barton and Jordan 2001). Graphic organizers can be used in a variety of settings. When used before reading, they help students link new information to their existing schema (Ausubel 1960). As a during-reading tool, graphic organizers assist students in taking notes and organizing information. Post-reading, they can be used to summarize and synthesize information or serve as a form of assessment. Novak (1991) found that when students constructed concept maps, their work better reflected their understanding of science concepts than traditional forms of testing.

## IDENTIFYING SIMILARITIES AND DIFFERENCES

Identifying similarities and differences is one of nine research-based strategies for increasing student achievement (Marzano, Pickering, and Pollock 2001). This strategy enhances students' understanding of concepts. You can help students identify similarities and differences by comparing, classifying, creating metaphors, and creating analogies (Marzano, Pickering, and Pollock 2001). Graphic organizers can be extremely useful in applying this strategy.

## INFOGRAPHICS

Information graphics, or *infographics,* are visual representations of data or information. Infographics are purposefully designed to make information easy to understand. They have four characteristics: (1) they present complex information in a quick and clear way; (2) they combine words and graphics to make meaning; (3) if done effectively, they are easier to understand than words alone; and (4) they are engaging and aesthetically appealing (Visually 2016).

Textbooks and nonfiction texts have included infographics for quite some time, but they are even more frequently found online. Although students may be drawn to them, they do need instruction and support in comprehending these complex visual elements. Asking students to create their own infographics often allows struggling writers to more effectively showcase their understanding.

## MAKING CONNECTIONS TO TEXT

Making connections is a reading comprehension strategy in which readers draw on their prior knowledge and experiences to relate to a text (Keene and Zimmerman 1997). When a reader makes connections to a text, he or she is engaged in the reading process and is actively thinking about the text. Thus, making connections can greatly enhance comprehension. Connections are classified into three categories: text-to-self, text-to-text, and text-to-world. Text-to-self connections link the reader's personal experiences to what is being read. Text-to-text connections link information from the current text to one that was previously read. Text-to-world connections link the text to phenomena that occur in the world at large but are not ones that the reader has necessarily experienced personally.

## MINI-LESSONS

A mini-lesson is a brief (5–15 minute) lesson with a narrow focus that can be used to teach a particular skill or strategy. Mini-lessons are best used when the skill or strategy is something that students will use often (Calkins 1986). Mini-lessons can be used with many types of groups, including individual students, small homogenous and heterogeneous groups, and the entire class. Typically, the structure of a mini-lesson is as follows: Teachers introduce the topic; demonstrate the strategy, skill, or concept; guide student practice; discuss the topic; volunteer more examples; and talk about what was taught. At the end of the mini-lesson, teachers give directions for the next activity (TeacherVision 2016).

# Appendix 2

Mini-lessons can be used to introduce a new skill or to provide additional support for students struggling with an existing skill. The activity following a mini-lesson should provide an immediate opportunity for students to apply and practice what they have learned.

## QUICK WRITES

A quick write is a literacy strategy that asks students to respond in 2–10 minutes to an open-ended question or prompt. Quick writes promote reflection and provide an opportunity for formative assessment. They can be used at a variety of times during a lesson: as a warm-up that either reviews content learned in the previous lesson or introduces a new topic and activates prior knowledge, during a lesson to help students connect new learning to prior knowledge, or at the end of a lesson to help students summarize and synthesize. Quick writes have been shown to promote understanding and are a valuable assessment tool (Green, Smith, and Brown 2007).

## RAFT PROMPTS

The RAFT strategy (Santa 1988) assists students in deepening their understanding of a topic through a focused writing assignment. Santa notes that when teachers compose content-area writing prompts, they are often too broad in scope; when students respond to these prompts, they do not explain clearly or completely. RAFT prompts, composed of four key ingredients (role of the writer, audience, format, and topic), alleviate both of those problems. Almost all RAFT prompts are written by a role other than the student, to an audience other than the teacher, and in a format other than the standard essay.

We've found that the specificity of these prompts support students as they communicate their understanding of a new topic. The prompts also require students to synthesize information, rather than just writing isolated facts (Barton and Jordan 2001). We've also found that using boldface type to present the elements of a RAFT helps students focus on the necessary components in their writing.

## READ-ALOUDS

Teacher read-alouds can be used for a variety of purposes, from introducing a text to the class as a whole to modeling fluent reading. Additionally, students of all ages simply enjoy being read to. A study of middle school students found that the second favorite in-class activity was read-aloud time (Ivey and Broaddus 2001). Research also indicates that if students hear books read aloud, they are more likely to select them for independent reading (Martinez et al. 1997). Reading aloud from a diversity of genres can thus help promote interest in informational and nonfiction text. Read-alouds can be teacher directed, in which the teacher drives the conversation about the text and asks questions to which the students respond. The teacher evaluates students' responses and continues to ask questions. An alternative approach is more collaborative, in which students are free to offer their own observations, ideas, and questions. Both types of read-alouds have their place in instruction, but teachers should carefully consider the purpose of the read-aloud as they determine which approach to use.

# Appendix 2

## SENTENCE STARTERS AND SENTENCE FRAMES

Sentence starters and sentence frames are effective techniques to assist students in developing proficiency with academic language (Herrmann 2016). Both strategies provide support for students as they speak or write about what they are learning but take slightly different forms. A sentence starter, as its name suggests, consists of the first few words of a sentence, such as "I noticed that ..." or "I claim that ... ." Students are responsible for completing the sentence. Sentence frames are very open-ended and are excellent for promoting observation during an investigation or reflection about what was observed. Sentence frames are more structured and provide a higher level of support for students. They consist of the majority of a sentence with key terms omitted. For example, a compare-and-contrast sentence frame might be "Both ________ and ________ have __________." Sentence frames allow students to use vocabulary from an investigation while providing a structure that they would have difficulty with on their own. Sentence frames are an effective scaffold when teaching a new text structure.

## SETTING A PURPOSE FOR READING

Setting a purpose for reading is a simple yet powerful strategy for promoting reading comprehension. In this strategy, the teacher or students identify a reason for reading a selection of text. For example, before reading a passage on cell organelles, a teacher might direct students to read to identify the function of the cell membrane. Setting a purpose can be as simple as a verbal cue to students before reading or as involved as creating an anticipation guide in which students rate their level of agreement or disagreement before and after reading. Establishing a purpose for reading allows students to discover why the reading assignment is valuable and assists in comprehension (Tovani 2000).

## SUMMARIZING TEXT

Summarizing is a reading strategy in which readers extract and paraphrase the essential ideas of a text. It requires readers to determine what is important, to condense this information, and to state it in their own words (Harvey and Goudvis 2007). For nonfiction and informational text, summarizing involves identifying the main ideas and supporting details. To be successful with this skill, students need explicit instruction and guided practice. The National Reading Panel found that providing instruction in summarizing helps students learn to identify main ideas, differentiate important and unimportant ideas, and remember what they read (NICHD National Reading Panel 2000).

## SYNTHESIZING INFORMATION

Synthesizing is a reading strategy in which a reader merges new information with prior knowledge to form a new idea, perspective, or opinion. It can also involve integrating information from multiple texts into a single idea. As with summarizing, students need explicit instruction in the strategy in addition to support as they practice the skill.

# Appendix 2

## TEACHING TEXT STRUCTURES

Nonfiction text is often organized in one of five common text structures: description, sequence, compare and contrast, cause and effect, and problem and solution (Oczkus 2014). Explicitly teaching text structures promotes comprehension of nonfiction text (Lipson 1996). You can teach text structures by introducing signal words and locating these words within nonfiction passages.

## THINK-PAIR-SHARE

Think-pair-share (Lyman 1987) is a three-step process that begins with students independently thinking about the prompt. Then, students pair up to discuss their thinking and identify the best response. The process concludes with each pair sharing its thinking with the whole group. Think-pair-share is widely applicable and has been found to increase student confidence and participation (Sampsel 2013).

## REFERENCES


Ausubel, D. P. 1980. The use of advanced organizers in the learning and retention of meaningful behavior. *Journal of Educational Psychology* 51 (5): 267–272.

Barton, M. L., and D. L. Jordan. 2001. *Teaching reading in science.* Aurora, CO: Mid-Continent Research for Education and Learning.

Calkins, L. 1986. *The art of teaching writing.* Portsmouth, NH: Heinemann.

Di Teodoro, S., S. Donders, J. Kemp-Davidson, P. Robertson, and L. Schuyler. 2011. Asking good questions: Promoting greater understanding of mathematics through purposeful teacher and student questioning. *The Canadian Journal of Action Research* 12 (2): 18–29.

Frayer, D. A., W. C. Frederick, and H. J. Klausmeier. 1969. A schema for testing the level of concept mastery. Technical Report No. 16, Research and Development Center for Cognitive Learning, University of Wisconsin, Madison.

Green, S. K., J. Smith III, and E. K. Brown. 2007. Using quick writes as a classroom assessment tool: Prospects and problems. *Journal of Educational Research & Policy Studies* 7 (2): 38–52.

Greenwood, S. 2004. Content matters: Building vocabulary and conceptual understanding in the subject areas. *Middle School Journal* 35 (3): 27–34.

Harvey, S., and A. Goudvis. 2007. *Strategies that work: Teaching comprehension for understanding and engagement.* 2nd ed. Portland, ME: Stenhouse Publishers.

Herrmann, E. 2016. Using sentence frames, sentence starters and signal words to improve language. MultiBriefs. *www.multibriefs.com/briefs/exclusive/using_sentence_frames.html#.V0Dr75ErLIU.*

Ivey, G., and K. Broaddus. 2001. Just plain reading: A survey of what makes students want to read in middle school classrooms. *Reading Research Quarterly* 36 (4): 350–371.

Keeley, P. 2014. *Science formative assessment: 50 more practical strategies for linking assessment, instruction, and learning.* Vol. 2. Thousand Oaks, CA: Corwin.

# Appendix 2


Keene, E. O., and S. Zimmerman. 1997. *Mosaic of thought: Teaching comprehension in a reader's workshop.* Portsmouth, NH: Heinemann.

Lipson, M. W. 1996. *Developing skills and strategies in an integrated literature-based reading program.* Boston, MA: Houghton-Mifflin.

Lyman, F. 1987. Think-pair-share: An expanding teaching technique. *MAA-CIE Cooperative News* 1 (1): 1–2.

Martinez, M., N. L. Roser, J. Worthy, S. Strecker, and P. Gough. 1997. Classroom libraries and children's book selections: Redefining "access" in self-selected reading. In *Inquires in literacy theory and practice: Forty-sixth yearbook of the National Reading Conference,* ed. C. K. Kinzer, K. A. Hinchman, and D. J. Leu, 265–272. Chicago: National Reading Conference.

Marzano, R. J., D. J. Pickering, and J. E. Pollock. 2001. *Classroom instruction that works: Research-based strategies for increasing student achievement.* Alexandria, VA: Association for Supervision and Curriculum Development.

McNeil, K. L., and J. S. Krajcik. 2011. *Supporting grade 5–8 students in constructing explanations in science: The Claim, Evidence, and Reasoning Framework for talk and writing*. Boston, MA: Pearson.

National Institute of Child Health and Human Development (NICHD) National Reading Panel. 2000. *Teaching children to read: An evidence-based assessment of the scientific research literature on reading and its implications for reading instruction.* Washington, DC: U.S. Government Printing Office.

Novak, J. D. 1991. Clarifying with concept maps. *The Science Teacher* 58 (7): 45–49.

Oczkus, L. 2014. *Just the facts! Close reading and comprehension of informational text.* Huntington Beach, CA: Shell Education.

Reinhartz, J. 2015. *Growing language through science, K–5: Strategies that work.* Thousand Oaks, CA: Corwin.

Sampsel, A. 2013. Finding the effects of think-pair-share on student confidence and participation. ScholarWorks. *http://scholarworks.bgsu.edu/honorsprojects/28.*

Santa, C. M. 1988. *Content reading including study systems.* Dubuque, IA: Kendall-Hunt.

TeacherVision. 2016. Focused mini-lessons. *www.teachervision.com/pro-dev/skill-builder/48710.html?page=1.*

Tovani, C. 2000. *I read it but I don't get it: Comprehension strategies for adolescent readers.* Portland, ME: Stenhouse Publishers.

Visually. 2016. What is an infographic? *https://visual.ly/m/what-is-an-infographic.*

Vlach, S., and J. Burcie. 2010. Narratives of the struggling reader. *The Reading Teacher* 63 (6): 522–525.

## Appendix 3

# Standards Alignment Matrices

### *NEXT GENERATION SCIENCE STANDARDS* (NGSS LEAD STATES 2013)

| Inquiry Units | Modeling Cells | The Genetic Game of Life | Seriously ... That's Where the Mass of a Tree Comes From? | Chemistry, Toys, and Accidental Inventions | Nature's Light Show: It's Magnetic! | Thermal Energy: An Ice Cube's Kryptonite! | Landfill Recovery | Sunlight and the Seasons | Getting to Know Geologic Time | The Toes and Teeth of Horses |
|---|---|---|---|---|---|---|---|---|---|---|
| **Disciplinary Core Ideas** | | | | | | | | | | |
| PS1.B: Chemical Reactions | | | | ✓ | | | | | | |
| PS2.B: Types of Interactions | | | | | ✓ | | | | | |
| PS3.A: Definitions of Energy | | | | | | ✓ | | | | |
| PS3.B: Conservation of Energy and Energy Transfer | | | | | | ✓ | | | | |
| PS3.D: Energy in Chemical Processes and Everyday Life | | | ✓ | | | | | | | |
| LS1.A: Structure and Function | ✓ | | | | | | | | | |
| LS1.B: Growth and Development of Organisms | | ✓ | | | | | | | | |
| LS1.C: Organization for Matter and Energy Flow in Organisms | | | ✓ | | | | | | | |
| LS3.A: Inheritance of Traits | | ✓ | | | | | | | | |
| LS3.B: Variation of Traits | | ✓ | | | | | | | | |
| LS4.A: Evidence of Common Ancestry and Diversity | | | | | | | | | | ✓ |
| ESS1.B: Earth and the Solar System | | | | | | | | ✓ | | |
| ESS1.C: The History of Planet Earth | | | | | | | | | ✓ | |
| ESS3.C: Human Impacts on Earth Systems | | | | | | | ✓ | | | |

# Appendix 3

| Inquiry Units | Modeling Cells | The Genetic Game of Life | Seriously … That's Where the Mass of a Tree Comes From? | Chemistry, Toys, and Accidental Inventions | Nature's Light Show: It's Magnetic! | Thermal Energy: An Ice Cube's Kryptonite! | Landfill Recovery | Sunlight and the Seasons | Getting to Know Geologic Time | The Toes and Teeth of Horses |
|---|---|---|---|---|---|---|---|---|---|---|
| **Science and Engineering Practices** | | | | | | | | | | |
| Asking questions (for science) and defining problems (for engineering) | | | | | | ✓ | ✓ | | | |
| Developing and using models | ✓ | ✓ | | | ✓ | | | | | |
| Planning and carrying out investigations | ✓ | ✓ | ✓ | ✓ | ✓ | ✓ | ✓ | ✓ | ✓ | ✓ |
| Analyzing and interpreting data | ✓ | ✓ | ✓ | ✓ | ✓ | ✓ | ✓ | ✓ | ✓ | ✓ |
| Using mathematics and computational thinking | | ✓ | ✓ | ✓ | | | | | | |
| Constructing explanations (for science) and designing solutions (for engineering) | ✓ | ✓ | ✓ | ✓ | ✓ | ✓ | ✓ | ✓ | ✓ | ✓ |
| Engaging in argument from evidence | | | | ✓ | | | ✓ | | | |
| Obtaining, evaluating, and communicating information | ✓ | ✓ | ✓ | ✓ | ✓ | ✓ | ✓ | ✓ | ✓ | ✓ |

# Appendix 3

| Inquiry Units | Modeling Cells | The Genetic Game of Life | Seriously … That's Where the Mass of a Tree Comes From? | Chemistry, Toys, and Accidental Inventions | Nature's Light Show: It's Magnetic! | Thermal Energy: An Ice Cube's Kryptonite! | Landfill Recovery | Sunlight and the Seasons | Getting to Know Geologic Time | The Toes and Teeth of Horses |
|---|---|---|---|---|---|---|---|---|---|---|
| **Crosscutting Concepts** | | | | | | | | | | |
| Patterns | | ✓ | | | | ✓ | | ✓ | | |
| Cause and effect: Mechanism and explanation | | ✓ | ✓ | | ✓ | ✓ | ✓ | ✓ | | ✓ |
| Scale, proportion, and quantity | | | | | | | | | ✓ | |
| Systems and system models | ✓ | | | | ✓ | | ✓ | ✓ | | ✓ |
| Energy and matter: Flows, cycles, and conservation | | | ✓ | ✓ | ✓ | ✓ | ✓ | | | |
| Structure and function | ✓ | ✓ | ✓ | | | | | | ✓ | |
| Stability and change | | ✓ | | ✓ | | | | | ✓ | ✓ |

# Appendix 3

## *COMMON CORE STATE STANDARDS, ENGLISH LANGUAGE ARTS* (NGAC AND CCSSO 2010)

| Inquiry Units | Modeling Cells | The Genetic Game of Life | Seriously ... That's Where the Mass of a Tree Comes From? | Chemistry, Toys, and Accidental Inventions | Nature's Light Show: It's Magnetic! | Thermal Energy: An Ice Cube's Kryptonite! | Landfill Recovery | Sunlight and the Seasons | Getting to Know Geologic Time | The Toes and Teeth of Horses |
|---|---|---|---|---|---|---|---|---|---|---|
| *CCSS.ELA-LITERACY.RH.6-8.2* Determine the central ideas or information of a primary or secondary source; provide an accurate summary of the source distinct from prior knowledge or opinions. | | ✓ | | | | | | | | |
| *CCSS.ELA-LITERACY.RST.6-8.1* Cite specific textual evidence to support analysis of science and technical texts. | ✓ | | ✓ | | | | ✓ | ✓ | | |
| *CCSS.ELA-LITERACY.RST.6-8.2* Determine the central ideas or conclusions of a text; provide an accurate summary of the text distinct from prior knowledge or opinions. | ✓ | | | | ✓ | | ✓ | ✓ | | |
| *CCSS.ELA-LITERACY.RST.6-8.3* Follow precisely a multistep procedure when carrying out experiments, taking measurements, or performing technical tasks. | ✓ | ✓ | ✓ | ✓ | ✓ | ✓ | | ✓ | | |
| *CSS.ELA-LITERACY.RST.6-8.4* Determine the meaning of symbols, key terms, and other domain-specific words and phrases as they are used in a specific scientific or technical context relevant to grades 6–8 texts and topics. | | | | | | ✓ | | | | |

# Appendix 3

| Inquiry Units | Modeling Cells | The Genetic Game of Life | Seriously ... That's Where the Mass of a Tree Comes From? | Chemistry, Toys, and Accidental Inventions | Nature's Light Show: It's Magnetic! | Thermal Energy: An Ice Cube's Kryptonite! | Landfill Recovery | Sunlight and the Seasons | Getting to Know Geologic Time | The Toes and Teeth of Horses |
|---|---|---|---|---|---|---|---|---|---|---|
| *CCSS.ELA-LITERACY.RST.6-8.6* Analyze the author's purpose in providing an explanation, describing a procedure, or discussing an experiment in a text. | | | | | | | | | | |
| *CCSS.ELA-LITERACY.RST.6-8.7* Integrate quantitative or technical information expressed in words in a text with a version of that information expressed visually (e.g., in a flowchart, diagram, model, graph, or table). | ✓ | | | | ✓ | ✓ | | ✓ | ✓ | |
| *CCSS.ELA-LITERACY.RST.6-8.8* Distinguish among facts, reasoned judgment based on research findings, and speculation in a text. | | | | | | | | | | |
| *CCSS.ELA-LITERACY.RST.6-8.9* Compare and contrast the information gained from experiments, simulations, videos, or multimedia sources with that gained from reading a text on the same topic. | ✓ | ✓ | ✓ | ✓ | ✓ | ✓ | ✓ | ✓ | ✓ | |
| *CCSS.ELA-LITERACY.WHST.6-8.1* Write arguments focused on discipline-specific content. | ✓ | | | | | | | ✓ | | |

# Appendix 3

| Inquiry Units | Modeling Cells | The Genetic Game of Life | Seriously ... That's Where the Mass of a Tree Comes From? | Chemistry, Toys, and Accidental Inventions | Nature's Light Show: It's Magnetic! | Thermal Energy: An Ice Cube's Kryptonite! | Landfill Recovery | Sunlight and the Seasons | Getting to Know Geologic Time | The Toes and Teeth of Horses |
|---|---|---|---|---|---|---|---|---|---|---|
| *CCSS.ELA-LITERACY.WHST.6-8.2* Write informative/explanatory texts, including the narration of historical events, scientific procedures/experiments, or technical processes. | ✓ | ✓ | | | | | ✓ | | ✓ | ✓ |
| *CCSS.ELA-LITERACY.WHST.6-8.4* Produce clear and coherent writing in which the development, organization, and style are appropriate to task, purpose, and audience. | | | | | | | | | ✓ | |
| *CCSS.ELA-LITERACY.WHST.6-8.6* Use technology, including the internet, to produce and publish writing and present the relationships between information and ideas clearly and efficiently. | | ✓ | | | | | | | | |
| *CCSS.ELA-LITERACY.WHST.6-8.7* Conduct short research projects to answer a question (including a self-generated question), drawing on several sources and generating additional related, focused questions that allow for multiple avenues of exploration. | | ✓ | | | | | | | | |

# Appendix 3

| Inquiry Units | Modeling Cells | The Genetic Game of Life | Seriously … That's Where the Mass of a Tree Comes From? | Chemistry, Toys, and Accidental Inventions | Nature's Light Show: It's Magnetic! | Thermal Energy: An Ice Cube's Kryptonite! | Landfill Recovery | Sunlight and the Seasons | Getting to Know Geologic Time | The Toes and Teeth of Horses |
|---|---|---|---|---|---|---|---|---|---|---|
| *CCSS.ELA-LITERACY.WHST.6-8.8* Gather relevant information from multiple print and digital sources, using search terms effectively; assess the credibility and accuracy of each source; and quote or paraphrase the data and conclusions of others while avoiding plagiarism and following a standard format for citation. | | ✓ | | | | | | | | |
| *CCSS.ELA-LITERACY.WHST.6-8.9* Draw evidence from informational texts to support analysis, reflection, and research. | ✓ | ✓ | | | ✓ | | ✓ | ✓ | ✓ | ✓ |

## REFERENCES


National Governors Association Center for Best Practices and Council of Chief State School Officers (NGAC and CCSSO). 2010. *Common core state standards.* Washington, DC: NGAC and CCSSO.

NGSS Lead States. 2013. *Next Generation Science Standards: For states, by states.* Washington, DC: National Academies Press. *www.nextgenscience.org/next-generation-science-standards.*

# Appendix 4

# Information for Safety Data Sheets

Safety data sheets (SDS) provide information about chemicals, such as hazards, composition, storage, and first aid measures in the event of skin or eye contact, swallowing, or inhalation of the chemical. We suggest printing the sheets for the chemicals you will be using and keeping them in an easily accessible location in your classroom. The sheets are arranged alphabetically within each chapter.

| Chemical | URL | QR Code |
|---|---|---|
| **Chapter 6: Modeling Cells** | | |
| Antacid tablets | *www.flinnsci.com/Documents/SDS/A/AntacidTablets.pdf* | QR CODE: Antacid Tablets |
| Dish soap | *www.flinnsci.com/Documents/SDS/C/CleanerHouseLiquidDish.pdf* | QR CODE: Dish Soap |
| Methylene blue stain | *www.flinnsci.com/Documents/SDS/M/MethyleneBlue.pdf* | QR CODE: Methylene Blue |
| Quieting solution | *www.flinnsci.com/Documents/SDS/OP/PolyvinylAlcoholSol.pdf* | QR CODE: Quieting Solution |

# Appendix 4

| Chemical | URL | QR Code |
|---|---|---|
| **Chapter 7: The Genetic Game of Life** | | |
| Rubbing alcohol | *www.flinnsci.com/Documents/SDS/IJ/IsopropylAlcohol.pdf* | QR CODE: Rubbing Alcohol |
| **Chapter 8: Seriously … That's Where the Mass of a Tree Comes From?** | | |
| Bromothymol blue | *www.flinnsci.com/Documents/MSDS/B/BromthymolBlueIndSol.pdf* | QR CODE: Bromothymol Blue |
| 1% sodium bicarbonate solution | *www.flinnsci.com/Documents/SDS/S/SodiumBicarbonateSol.pdf* | QR CODE: 1% Sodium Bicarbonate Solution |
| **Chapter 9: Chemistry, Toys, and Accidental Inventions** | | |
| Baby powder | *www.lakeland.edu/PDFs/MSDS/1395/Talcum%20Powder%20 %28Commerical%29.pdf* | QR CODE: Baby Powder |
| Baking soda | *www.flinnsci.com/Documents/SDS/S/SodiumBicarbonate.pdf* | QR CODE: Baking Soda |

# Appendix 4

| Chemical | URL | QR Code |
|---|---|---|
| Borax | *www.omsi.edu/sites/all/FTP/files/kids/Borax-msds.pdf* | QR CODE: Borax |
| Boric acid solution | *www.flinnsci.com/Documents/SDS/B/BoricAcidSol.pdf* | QR CODE: Boric Acid Solution |
| Castile soap | *www.sciencelab.com/msds.php?msdsId=9923335* | QR CODE: Castile Soap |
| Laundry detergent | *www.iaprisonind.com/downloads/msds/IPI-AllTemp.pdf* | QR CODE: Laundry Detergent |
| Liquid starch | *www.lakeland.edu/PDFs/MSDS/840/Sta-Flo%20Spray%20Starch%20 %28Dial%20Corp%29.pdf* | QR CODE: Liquid Starch |
| Silicone oil | *www.flinnsci.com/Documents/SDS/S/SiliconeOil.pdf* | QR CODE: Silicone Oil |

# Appendix 4

| Chemical | URL | QR Code |
|---|---|---|
| Vinegar | *www.flinnsci.com/Documents/SDS/UV/Vinegar.pdf* | QR CODE: Vinegar |
| White glue | *www.flinnsci.com/Documents/SDS/G/GlueWhite.pdf* | QR CODE: White Glue |
| **Chapter 10: Nature's Light Show: It's Magnetic!** | | |
| Iron filings | *www.flinnsci.com/Documents/SDS/IJ/IronPowder.pdf* | QR CODE: Iron Filings |
| **Chapter 12: Landfill Recovery** | | |
| Iron filings | *www.flinnsci.com/Documents/SDS/IJ/IronPowder.pdf* | QR CODE: Iron Filings |
| **Chapter 14: Getting to Know Geologic Time** | | |
| Plaster of paris | *www.flinnsci.com/Documents/MSDS/C/CalciumSulfate.pdf* | QR CODE: Plaster of Paris |

# Image Credits

## CHAPTER 1

Figure 1.1: NSTA Press

## CHAPTER 6

p. 78: Mariana R. Villarreal, public domain.
*https://commons.wikimedia.org/wiki/File:Diagram_cell_nucleus_no_text.svg.*

p. 79: Mariana R. Villarreal, public domain.
*https://commons.wikimedia.org/wiki/File:Mitosis_cells_sequence.svg.*

p. 83: Mariana R. Villarreal, public domain.
*https://commons.wikimedia.org/wiki/File:Animal_mitochondrion_diagram_unlabelled.svg.*

p. 84: NSTA Press

p. 87: Mariana R. Villarreal, public domain.
*https://commons.wikimedia.org/wiki/File:Cell_membrane_detailed_diagram_blank.svg.*

## CHAPTER 8

pp. 161–162: Terry Shiverdecker using Pixton Comics Inc. services available at *www.pixton.com.*

## CHAPTER 10

p. 199: NSTA Press

## CHAPTER 12

Figure 12.1: NSTA Press

## CHAPTER 14

Figure 14.1: Roger Steinberg, public domain.
*http://serc.carleton.edu/NAGTWorkshops/time/visualizations_teachtips/60786.html.*

## CHAPTER 15

p. 324: NSTA Press

pp. 325–326: Mcy jerry, CC BY-SA 3.0. *https://commons.wikimedia.org/w/index.php?curid=496577.*

# INDEX

Page numbers in **boldface** type refer to tables or figures.

# Index

# Index

# Index

# Index

# Index

# Index

# Index

# Index

# Index

# Index

# Index

# Index

# Index